Eduardo Pagel Floriano

Forest Inventory

Eduardo Pagel Floriano

Forest Inventory

Forest Measurement and Monitoring

ScienciaScripts

Imprint
Any brand names and product names mentioned in this book are subject to trademark, brand or patent protection and are trademarks or registered trademarks of their respective holders. The use of brand names, product names, common names, trade names, product descriptions etc. even without a particular marking in this work is in no way to be construed to mean that such names may be regarded as unrestricted in respect of trademark and brand protection legislation and could thus be used by anyone.

Cover image: www.ingimage.com

This book is a translation from the original published under ISBN 978-3-639-68325-7.

Publisher:
Sciencia Scripts
is a trademark of
Dodo Books Indian Ocean Ltd. and OmniScriptum S.R.L publishing group

120 High Road, East Finchley, London, N2 9ED, United Kingdom
Str. Armeneasca 28/1, office 1, Chisinau MD-2012, Republic of Moldova, Europe
Printed at: see last page
ISBN: 978-620-8-28127-4

EDUARDO PAGEL FLORIANO

Forest Inventory

Rio Largo, AL, Brazil
2024

SUMMARY

PRESENTATION

This compendium was initially developed as a text for use in the Forest Inventory subject of the Forest Engineering course at the Engineering and Agrarian Sciences Campus of the Federal University of Alagoas. The aim was to create a didactic text as comprehensible as possible for students and professionals in the field. It was then improved and a second version published online.

This edition was motivated by the large number of accesses and downloads by people from various countries. In addition to revising the text, some important topics have been added that had not been covered previously, and the SAS programs for analyzing forest inventory data have been retained.

An example of a continuous forest inventory with partial repetition by stratified random sampling was also added, with estimates of the data from the trees in the plots that were not measured, according to the methodology developed by the author over the years working with forest inventories and teaching the subject.

We hope this text will be useful to anyone who needs it.

Maceió, October 2, 2024.

Eduardo Pagel Floriano

1. INTRODUCTION

The Forest Inventory subject in Forest Engineering courses aims to teach students how to analyze forest populations in quantitative, qualitative and dynamic terms, based on dendrometric techniques and statistical principles, in order to carry out forest inventories with an emphasis on forest administration and management.

Forest Inventory can be conceptualized as the quantitative and sometimes qualitative assessment of forest resources in a given area with forest cover.

The technologies used in forest inventories have evolved since the 18th century, when visual assessments of forests were made in the following sequence:

- 18th century - In Europe, quantification was carried out visually and forest areas were divided into smaller units to facilitate estimates;
- 19th century - dendrometric and statistical sampling techniques and relationships between d, h and v were used to make estimates;
- 20th century - from the 1960s onwards, continuous inventories became more frequent and computerized;
- 21st century - The use of images, drones with laser trackers and artificial intelligence are enabling the development of technologies for carrying out remote inventories.

1.1 Importance of forest inventories

The main items of importance in forest inventories are listed below:

- To provide information on stocks and trends for the formation of public policies for the conservation and supply of forest products;
- Monitoring the evolution of cultivated forests and their wood stocks over time;
- Provide information for planning the production of forest products;
- Monitoring the evolution of protection and conservation forests in order to plan actions for their maintenance;
- Monitoring urban trees and forests in order to provide information for planning.

1.2 Types of forest inventories

Forest inventories can be classified as follows:

- Spatial scope: national, state, regional, local;
- At the administrative or planning level: strategic, tactical, operational;
- The form of the survey: census (100%) or sampling (<100%);
- The time approach: temporary or continuous;
- To the objective: wood, biomass, growth, competition, management, phytosociology, ecology, afforestation.
- Level of detail: detailed, reconnaissance, exploratory.

1.2.1 Classification according to the object of the inventory

Forest inventories can be

- Natural forest managed for the production of forestry goods;
- Natural forest for protection or conservation;
- Forest planted for restoration;
- Planted production forest;
- Urban forest;
- Urban afforestation.

1.2.2 Environment of the inventory site and surroundings

It is important to characterize the environment and surroundings of the forest areas that are to be the subject of the forest inventory in order to facilitate planning and serve as a reference for carrying it out, including:

- Mapping: forest areas, area to be inventoried and type of forest, reserve and preservation areas, bodies of water, access roads and infrastructure, etc;
- Soils;
- Climate;
- Type of environment: rural; urban.
- Conservation units;
- Nearest towns and population;
- Resources and places of support: pharmacies, markets, restaurants, hospitals and health centers, accommodation, police stations, repair shops, etc.

1.2.3 Level of Detail and Accuracy required

The level of detail of a forest inventory is directly related to its objective and the criticality of the forest involved.

- The level of detail depends:
  - Objective and financial resources;
  - From criticality to defining precision:
  - Legislation and the requirements of licensing bodies;
  - The risks arising from errors;
- Required accuracy can be:
  - = 100%: census;
  - < 100%: sample;
- The detail of the inventory is characterized by the sampling intensity, which depends on the maximum sampling error allowed (LE) in relation to the average;
- To determine the sampling intensity, it is necessary to
  - Establish the probability of confidence for the mean (P) and the error limit in relation to the mean (LE%=200 (1-P)); or p = LE% / 2, where p is the probability of error;
  - Determine the mean ($\bar{x}$) and variance (s^2) using a pilot inventory and the absolute sampling error limit (LE = LE% . $\bar{x}$).

A pilot or preliminary inventory is the sampling of a small number of sample units in order to determine the preliminary mean and variance for the population, which will allow the total number of sample units needed to be determined for the maximum limit of error in relation to the mean (LE%) established.

1.2.4 Time approach

Inventories can be carried out on a single occasion to establish the current situation of the forest and its stocks, or on multiple occasions to monitor its growth and evolution, and are said to be temporary and continuous, respectively. The approach over time can be as follows:

- Temporary Forest Inventories
  - Temporary sampling units;
  - Current situation;
  - Only once;
  - Sporadic;
  - Short-term planning.
- Continuous Forest Inventories
  - Permanent sampling units;
  - Multiple correlated occasions;
  - Growth;
  - Evolution;
  - Medium and long-term planning.

1.3 Pilot inventory

Usually, a forest inventory begins by sampling a few plots, identical to the plots in the definitive inventory, in order to determine how many sampling units will be needed in the definitive inventory. The number of sampling units in the pilot inventory should be such that an approximation of the population's mean and variance is known. The more heterogeneous the population, the more units should make up the pilot sample.

The pilot inventory has the following objectives:

- determine the preliminary mean and variance;
- test the chosen methodology;
- determine the time required to move, mark and measure the sampling units;
- define the equipment and materials needed;
- train the inventory team;
- determine the number of sampling units needed in the final inventory.

The methodology to be used in the pilot inventory must be the same as that planned for the definitive inventory.

In the pilot inventory, travel time must be measured, and the trees in the sample plots must be marked and data collected.

After the pilot inventory and with the data obtained, the definitive inventory is planned.

The number of sampling units to be used in the pilot inventory should always be more than 5 and should be greater the larger and more heterogeneous the population. The number of sample plots (n) in the pilot inventory, for each stratum or management unit, or forest stand, can be indicated by the equation:

$$n = a + b\,A^{c}$$

Where: n = number of sampling units recommended for the pilot inventory; a, b, c = coefficients of the equation that may vary depending on the homogeneity of the forest; A = area of the forest stand considered (stratum, or management unit, or forest stand plot).

For populations with medium to high homogeneity, the coefficients of the equation can be a = 3.075, b = 0.681 and c = 0.644, and the equation is as follows:

$$n = 3.075 + 0.681\,A^{0,644}$$

If the population is heterogeneous, the b coefficient can be increased to 1 and the equation is reduced to:

$$n = 3.075 + A^{0,644}$$

If the population is very homogeneous, the coefficient "b" can be reduced to 0.4, the coefficient "a" increased to 4 and the equation is as follows:

$$n = 4 + 0.4\,A^{0,644}$$

1.4 Sampling Method, System and Process

A forest inventory is characterized by the method, system and sampling process used, which will depend mainly on the type of forest, the size of the forest and ease of access, as follows:

- Sampling method:
  - Fixed area sampling units: square, rectangular, circular;
  - Variable area units: Bitterlich sampling, Strand sampling, quadrants, nearest tree, etc;
- Sampling system:
  - Random (with or without restrictions) → emphasis on variance
  - Systematic (with or without restrictions) → emphasis on the average
- The main sampling processes:
  - AAS - Simple Random Sampling;

- SEA - Stratified Random Sampling;
- AS - Systematic sampling;
    - Methods
        - In rows - the plots in the rows are spaced apart from each other;
        - In strips - the plots in the rows have no spaces between them;
        - In grids - the plots are located at the intersection of equidistant horizontal and vertical lines;
    - Systems
        - AS - With a single random start
        - AS - With multiple random starts (Figura 1);
- AC - Conglomerate Sampling;
- AME - Multiple Stage Sampling.

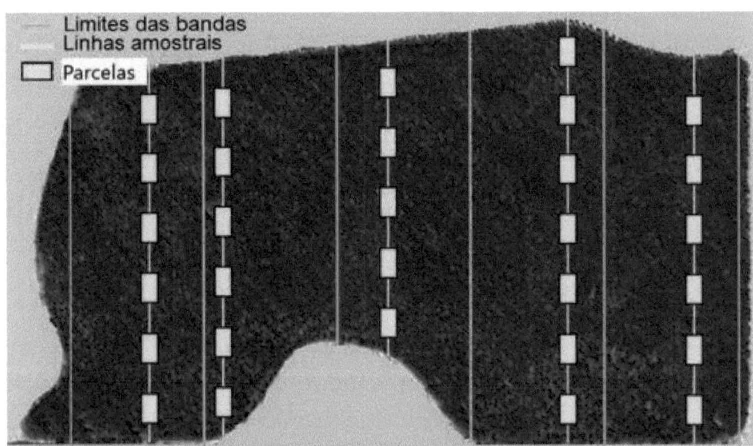

FIGURA 1 - Forest inventory using the line sampling method, with multiple random starts.

2. SAMPLING THEORY

In order to get to know a forest population or community, we need data about it. The accuracy of the data on the attributes and variables surveyed depends mainly on the objectives of the forest, forest typology, dimensions and ease of access to the forest area subject to the inventory. The more critical the objective of the forest, the greater the precision required in data collection.

Forests have thousands or even millions of individuals and it is very expensive to measure them all, so sampling is usually used. Sampling+ is the assessment of a population or community from a fraction of it.

The main characteristics of the inventories and forests that must be established are:

- Sample variables and attributes;
- Sample statistics;
- Sampling intensity;
- Type of units and sampling methods;
- Sampling systems and processes.

2.1 Variables

Variables are the characteristics of individuals in the population that can be measured or parameterized. Variables can be of two classes: quantitative and qualitative (Figura 2). Quantitative variables can be of two types: discrete, which assume only integer values; and continuous, which assume real values. Qualitative variables are also divided into two classes:

nominal, which are not ordered, such as species, health, etc.; and ordinal, which can be ordered, such as age, position in the vertical stratum of the forest, successional stage of the vegetation, stem quality class.

Examples:

- Measured variable: tree diameter at 1.3 m height;
- Parameterized variable: average spacing between trees (MS) calculated by:

$$MS = \sqrt{(10000m^2 / N)}$$

Where: MS = average spacing between trees in meters; N = number of trees per hectare.

The main dendrometric variables and tree attributes obtained from forest inventories :

- Tree species;
- diameter (d) ;
- bark thickness (e);
- height (h); basal area (g);
- form factor (f);
- volume (v);
- shelled volume (vs);
- Crown diameter (CD);
- canopy base height (h_{bc});
- canopy projection surface (SPC);
- Stem height (hf);
- commercial volume (vc);
- Stem quality (QF);
- Wooden assortments.

Population variables obtained from timber forest inventories are:

- Location (coordinates) and area;
- age (t);
- average diameter ($\bar{d}$);
- average basal area ($\bar{g}$);
- average height ($\bar{h}$);
- number of trees per hectare (N);
- basal area per hectare (G);
- volume per hectare (V);
- bark-free volume per hectare (V_s).

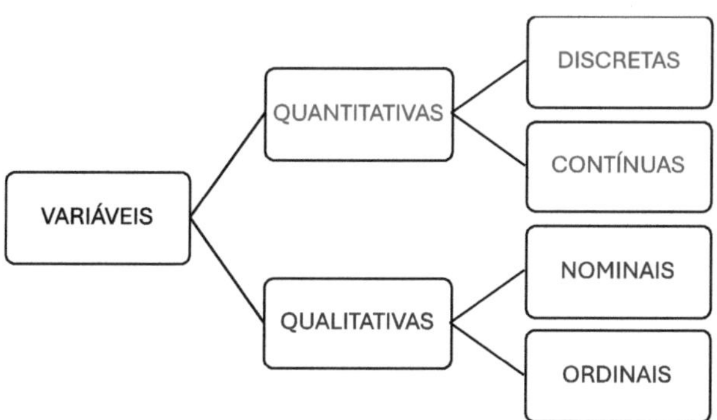

FIGURA 2 - Types of statistical variables.

2.2 Attributes obtained from forest inventories:

These are the qualitative characteristics of the individuals in the population or community being studied. The statistics carried out on them are called attribute statistics. They are enumerated in the survey, i.e. they are counted, as they are not measurable.

Examples of attributes (qualitative variables):

- number of individuals attacked by diseases in the population;
- Number of trees per hectare (N);
- Class of quality or utilization of the stem (example of ordinal attribute):
  - 1 - from 0 to 25%;
  - 2 - from 26 to 50%;
  - 3 - from 51 to 75%;
  - 4 - from 75 to 100%;
- Tree conditions (example of a nominal attribute):
  - normal - N;
  - dead - M;
  - bifurcated - B;
  - sick/attacked - D;
  - broken - Q;
  - inclined - I;
  - others - X.

2.3 Symbology and Nomenclature

The symbols and names used in this text are described below:

- y = dendrometric variable measured in a sample;
- w = weight of a sampling level;
- h (1; 2; ...; H) = index of the 1st level units;
- H = number of 1st level units in the sample;
- N_H = potential number of 1st level units in the population;

- i (1; 2; ...; n_h) = index of 2nd level units within 1st level units;
- h (1; 2; ...; m_{hi}) = index of 3rd level unit within 2nd level units, within 3rd level units;
- n = total number of observations in the sample;
- N = potential number of sample units in the population;
- y_{hij} = observed value of the variable in unit h in group i of stratum h;
- w_{hij} = sampling weight for unit h in group i of stratum h;
- $f_h = H / N_H$ = sampling rate of the 1st level units, which is used in estimating the variance of the Taylor series, is the fraction of the 1st level units (first stage) selected for the sample;
- Limit of sampling error (LE, LE%) = maximum error allowed for sampling:

$$LE = LE\% \,.\, \bar{y} \,/\, 100, \qquad LE\% = 100 \,.\, LE \,/\, \bar{y}$$

$$LE\% = 100\,(2\,.\,\alpha)$$

Where: α = 1 - P; P = probability of confidence for the mean.

- Probability of confidence for the mean (P)

$$P = 1 - \alpha$$

2.4 Main sample statistics

The main sample statistics determined in forest inventories are listed below:

2.4.1 Sample mean ($\bar{y}$):

$$\bar{y} = \frac{\sum_{h=1}^{H}\sum_{i=1}^{n_h}\sum_{j=1}^{m_{hi}} w_{hij}\ y_{hij}}{\sum_{h=1}^{H}\sum_{i=1}^{n_h}\sum_{j=1}^{m_{hi}} w_{hij}}$$

2.4.2 Median (Med):

It is the element that divides a set of elements into two parts with the same number of elements.

Example: in the set {5 6 8 9 10 12 13}, the median is 9.

2.4.3 Fashion (Mo):

Mode is the most frequent element in a set of elements.

Example: in the set {5 6 8 9 9 10 12 13}, the mode is 9.

2.4.4 Lower Limit (LI), Upper Limit (LS) and Amplitude (A):

The lower limit of a set of elements is the lowest value found and the upper limit is the highest value, while the amplitude is the difference between the two.

Example: in the set {5 6 8 9 10 12 13}, the smallest value is 5 and the largest is 13, so the amplitude is calculated by A = LS - LI = 13 - 5 = 8.

2.4.5 Sample variance (s^2):

$$s^2 = \frac{\sum_{h=1}^{H}(y_h - \bar{y})^2}{H-1}$$

2.4.6 Sample standard deviation (s):

$$s = \sqrt{s^2}$$

2.4.7 Coefficient of Variation (CV):

$$CV = 100 . s / \bar{y}$$

2.4.8 Variance of the sample mean

The Taylor series method for estimating the variance of the mean is calculated as follows (SAS, 2016):

$$\hat{V}(\hat{\bar{Y}}) = \sum_{h=1}^{H} \hat{V}_h(\hat{\bar{Y}})$$

Onde, se $n_h > 1$, então

$$\hat{V}_h(\hat{\bar{Y}}) = \frac{n_h(1 - f_h)}{n_h - 1} \sum_{i=1}^{n_h} (e_{hi\cdot} - \bar{e}_{h\cdot\cdot})^2$$

$$e_{hi\cdot} = \left(\sum_{j=1}^{m_{hi}} w_{hij} \, (y_{hij} - \hat{\bar{Y}}) \right) / \, w_{\cdots}$$

$$\bar{e}_{h\cdot\cdot} = \left(\sum_{i=1}^{n_h} e_{hi\cdot} \right) / \, n_h$$

E, se $n_h = 1$, então

$$\hat{V}_h(\hat{\bar{Y}}) = \begin{cases} \text{faltando} & \text{se } n_{h'} = 1 \text{ para } \quad h' = 1, 2, \ldots, H \\ 0 & \text{se } n_{h'} > 1 \text{ para algum } \quad 1 \leq h' \leq H \end{cases}$$

2.4.9 Standard Error of the Sample Mean ($S_{\bar{y}}$):

$$S_{\bar{y}} = \sqrt{S_{\bar{y}}^2}$$

2.4.10 Absolute sampling error (Ea):

$$Ea = \pm t_{(p;\, GL)} . S_{\bar{y}}$$

2.4.11 Relative sampling error (Er):

$$Er = \pm 100 . Ea / \bar{y}$$

2.4.12 Lower limit of the average (LI):

$$LI_m = \bar{y} - Ea$$

2.4.13 Upper limit of the mean (LS):m

$$LS_m = \bar{y} + Ea$$

2.4.14 Minimum confidence estimate (MCE):

$$EMC = LI_m$$

2.4.15 Confidence Interval (CI):

$$IC\,[\,(LI_m) \leq \mu \leq (LS_m)\,] = P$$

2.5 Student's t distribution

Sampling theory is based on probability theory and uses Student's t-distribution to determine the necessary sample size to represent the population with a given probability of confidence for the mean (P). The t-distribution is a statistical probability distribution developed by "Student", the pseudonym of William Sealy Gosset, an English chemist and statistician (July 13, 1876 - October 16, 1937).

The t-distribution is theoretical, symmetrical and similar to the normal distribution curve with wider tails, the number of degrees of freedom (v) being the only parameter that defines it.

If Z and V are two independent random variables, where Z has a standard normal distribution with mean 0 and standard deviation 1 and where V has a Chi-squared distribution with v degrees of freedom, the distribution of a random variable t will be given by:

$$t = \frac{Z}{\sqrt{V/_v}}$$

The probability density function of the t distribution is:

$$f(t) = \frac{\Gamma(\frac{\nu+1}{2})}{\sqrt{\nu\pi}\,\Gamma(\frac{\nu}{2})}\left(1+\frac{t^2}{\nu}\right)^{-(\frac{\nu+1}{2})}$$

Where: f(t) = probability density of t; Γ = Gamma function; v = degrees of freedom.

The one-tailed and two-tailed t-distribution, with z-values, is shown in Tabela 1 below.

TABELA 1 - One- and two-tailed t-distribution

	Prob. accum.	$t_{0,75}$	$t_{0,80}$	$t_{0,85}$	$t_{0,90}$	$t_{0,95}$	$t_{0,975}$	$t_{0,9875}$	$t_{0,995}$
	Unicaudal	0,2500	0,2000	0,1500	0,1000	0,0500	0,0250	0,0125	0,0050
	Two-tailed	0,5000	0,4000	0,3000	0,2000	0,1000	0,0500	0,0250	0,0100
Degrees of Freedom	1	1,0000	1,3764	1,9626	3,0777	6,3138	12,7062	25,4517	63,6567
	2	0,8165	1,0607	1,3862	1,8856	2,9200	4,3027	6,2053	9,9248
	3	0,7649	0,9785	1,2498	1,6377	2,3534	3,1824	4,1765	5,8409
	4	0,7407	0,9410	1,1896	1,5332	2,1318	2,7764	3,4954	4,6041
	5	0,7267	0,9195	1,1558	1,4759	2,0150	2,5706	3,1634	4,0321
	6	0,7176	0,9057	1,1342	1,4398	1,9432	2,4469	2,9687	3,7074
	7	0,7111	0,8960	1,1192	1,4149	1,8946	2,3646	2,8412	3,4995
	8	0,7064	0,8889	1,1081	1,3968	1,8595	2,3060	2,7515	3,3554
	9	0,7027	0,8834	1,0997	1,3830	1,8331	2,2622	2,6850	3,2498
	10	0,6998	0,8791	1,0931	1,3722	1,8125	2,2281	2,6338	3,1693
	11	0,6974	0,8755	1,0877	1,3634	1,7959	2,2010	2,5931	3,1058
	12	0,6955	0,8726	1,0832	1,3562	1,7823	2,1788	2,5600	3,0545
	13	0,6938	0,8702	1,0795	1,3502	1,7709	2,1604	2,5326	3,0123
	14	0,6924	0,8681	1,0763	1,3450	1,7613	2,1448	2,5096	2,9768
	15	0,6912	0,8662	1,0735	1,3406	1,7531	2,1314	2,4899	2,9467
	16	0,6901	0,8647	1,0711	1,3368	1,7459	2,1199	2,4729	2,9208
	17	0,6892	0,8633	1,0690	1,3334	1,7396	2,1098	2,4581	2,8982
	18	0,6884	0,8620	1,0672	1,3304	1,7341	2,1009	2,4450	2,8784
	19	0,6876	0,8610	1,0655	1,3277	1,7291	2,0930	2,4334	2,8609
	20	0,6870	0,8600	1,0640	1,3253	1,7247	2,0860	2,4231	2,8453
	21	0,6864	0,8591	1,0627	1,3232	1,7207	2,0796	2,4138	2,8314
	22	0,6858	0,8583	1,0614	1,3212	1,7171	2,0739	2,4055	2,8188
	23	0,6853	0,8575	1,0603	1,3195	1,7139	2,0687	2,3979	2,8073
	24	0,6848	0,8569	1,0593	1,3178	1,7109	2,0639	2,3909	2,7969
	25	0,6844	0,8562	1,0584	1,3163	1,7081	2,0595	2,3846	2,7874
	26	0,6840	0,8557	1,0575	1,3150	1,7056	2,0555	2,3788	2,7787
	27	0,6837	0,8551	1,0567	1,3137	1,7033	2,0518	2,3734	2,7707
	28	0,6834	0,8546	1,0560	1,3125	1,7011	2,0484	2,3685	2,7633
	29	0,6830	0,8542	1,0553	1,3114	1,6991	2,0452	2,3638	2,7564
	30	0,6828	0,8538	1,0547	1,3104	1,6973	2,0423	2,3596	2,7500
	40	0,6807	0,8507	1,0500	1,3031	1,6839	2,0211	2,3289	2,7045
	60	0,6786	0,8477	1,0455	1,2958	1,6706	2,0003	2,2990	2,6603
	80	0,6776	0,8461	1,0432	1,2922	1,6641	1,9901	2,2844	2,6387
	100	0,6770	0,8452	1,0418	1,2901	1,6602	1,9840	2,2757	2,6259
	500	0,6750	0,8423	1,0375	1,2832	1,6479	1,9647	2,2482	2,5857
	1000	0,6747	0,8420	1,0370	1,2824	1,6464	1,9623	2,2448	2,5808
z		**0,674**	**0,842**	**1,036**	**1,282**	**1,645**	**1,960**	**2,326**	**2,576**
P		**50,0%**	**60,0%**	**70,0%**	**80,0%**	**90,0%**	**95,0%**	**97,5%**	**99,0%**

Where: P = probability of confidence for the mean. If the error limit is 10%, the probability of confidence for the mean is 95%. Example: alpha = 5%; P = 95%; Limit of Error (2*alpha) LE = 10%.

The value of t can be found using the formula:

$$t = \frac{\bar{X} - \mu}{\sqrt{\frac{s^2}{n}}} \quad ,ou \quad \mu = \bar{x} - t\sqrt{\frac{s^2}{n}} \quad ,ou\ ainda \quad \mu = \bar{x} \pm t\left(\frac{s}{\sqrt{n}}\right)$$

Where: t = calculated value of t; $\bar{x}$ = sample mean; μ = population mean; s^2 = sample variance; s = sample standard deviation; n = number of sample units; $\pm \mathrm{t}\left(\frac{\mathrm{s}}{\sqrt{\mathrm{n}}}\right)$ = sampling error.

The t-value used is two-tailed to define the lower and upper limits $[\bar{x} - t\,(s/\sqrt{n})]$ and upper $[\bar{x} + t\,(s/\sqrt{n})]$ of the mean and can be obtained from t-tables with probability p and n-1 degrees of freedom or from the INV.T.BC(p; n-1) function in LibreOffice Calc v7.1 and MS-Excel 365, or from the INVT(p; n-1) function in older versions of the two software packages.

Care should be taken, as the INV.T(p; n-1) function in both software packages returns the one-tailed inverse of the t-distribution and should not be used to determine the number of sampling units required or to determine the sampling error.

Example of calculating the sampling error for more or less than the mean:

- Probability of confidence for the mean P= 95%, so the error limit is 10%, i.e. $p = \pm 5\%$;
- Sample units = 10, so the degrees of freedom are 9;
- The inverse function of two-tailed t in the spreadsheet is written as:

=INVT(1-0.95; 9), or

=INVT(0.05; 9), or

=INV.T.BC(1-0.95; 9), or even

=INV.T.BC(0.05; 9)

Where the return of the function is the value of t = 2.262;

- The average volume $\bar{V}$ = 200 m³/ha, and the variance of the mean $S_{\bar{x}}^2$ = 640 (m³/ha)² the sampling error will be calculated as:

$$E = t\sqrt{\frac{S_{\bar{x}}^2}{n}} = E = 2{,}262\sqrt{\frac{640}{10}} = +/- \ 18.096 m^3/ha, \ or$$

$$E\% = 100 \ . \ 18{,}096 \ / \ 200 = +/- \ 9{,}048\%$$

2.6 Finite and infinite populations

The size of the sample in relation to the size of the population defines whether our population is finite or infinite. A population is considered infinite if the sample is less than 2% of the population.

Example: if all the sampling units add up to 2 hectares and the population is 100 hectares, the sampling intensity is 2%, so it is ≥2% and the population is considered finite; the formulas for statistical calculations need to be given a correction factor (CF) for finite populations in order to eliminate sampling bias.

2.7 Sample unit

A sampling unit is a small portion of the population where measurements are made of the variables and attributes of the individuals and the population, of sufficient size to contain the variability of the individuals at the site. The sampling units must be sufficient in number to cover the population variation and allow estimates of population parameters to be made within the precision required for sampling. The sampling error varies according to the size and number of sampling units, as in Figura 3.

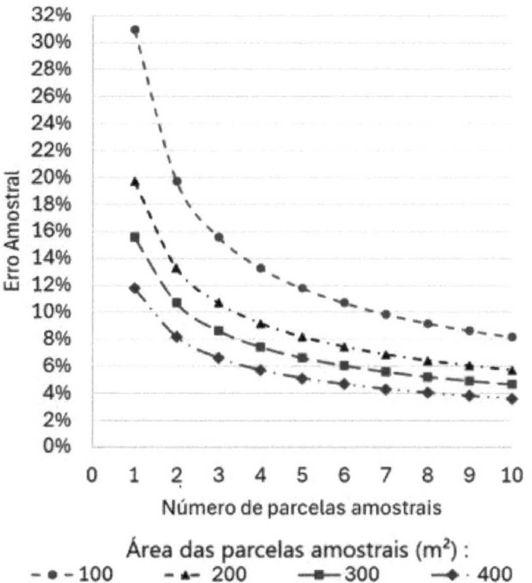

FIGURA 3 - Behavior of the sampling error in relation to the size of the sampling unit and the number of sampling units.

The types of sampling units define the sampling methods, which can be fixed-area sampling units or variable-area units, with different subtypes.

The influence of the shape and size of the sampling units is as follows:

- Shape:
  - Homogeneous population - larger area and smaller perimeter = shorter measurement time;
  - Heterogeneous population - higher length/side ratio reduces variance between.
- Size:

- The larger the dimensions, especially the length, the greater the variance within the unit;
- Ideal of 20 to 100 individuals to maximize variance within and minimize variance between AUs in order to have the lowest sampling error (Figura 3).

2.8 Types and demarcation of sampling units

The sampling units in forests, commonly called plots, can be of three types:

- Punctual - these are sampling units where only the center of the unit is fixed; the areas of the sampling units vary from one to another; the main types are the Bitterlich, Strand, Prodan, Quadrant and Nearest Tree sampling units.
- Linear - these are sampling units with no fixed area, as the length of the units varies from one to another, based on lines or transects; the most common types are: lines, transects and nearest neighbor;
- Fixed Area - these are sampling units whose shape and size is standardized for all plots; the main types are: square, rectangular and circular.

2.8.1 Measurements and estimates in sampling units

In any case, the diameters of all the trees in the plots are measured and at least a percentage of the trees have their heights measured in order to then adjust hypsometric relationship equations and estimate the heights of trees whose heights have not been measured, or all heights are measured, which makes the inventory more expensive.

The frequency of trees in the plots is calculated by counting the number of trees in the plot. The average diameter and average basal area of the sampling unit is calculated by the arithmetic mean of all the trees in the plot, except the dead ones. The average height is calculated from the trees whose heights were measured and estimated, except for dead trees. Tree volume should be estimated using a volumetric equation and its average calculated as the arithmetic mean of all the trees in the plots, except the dead ones. The averages per hectare for tree frequency (N), basal area (G) and volume (V) are found by multiplying the Area Proportionality Factor (F), which is calculated by the ratio between the area of 1 ha and the area of the sampling unit. The Area Proportionality Factor is calculated using the equation:

$$F = A / a$$

Where: F = Area Proportionality Factor; A = 10000 m^2 (=1 ha); a = area of the sampling unit in m^2.

The areas of the sampling units (a) are determined according to the type of sampling unit as described in "Chapter 3 Sampling Methods".

2.8.2 Square or rectangular UA demarcation

The size of the area and the sides are defined and the center of the AU is demarcated on the access line (in black in Figure 4). Figura 4).

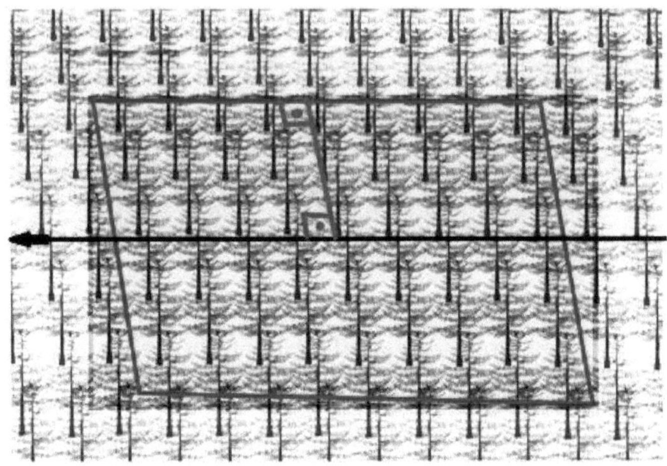

FIGURA 4 - Square or rectangular UA demarcation.

A right angle is formed from the center to half of the sides and from the middle of the side to the corners, forming a cross. You should demarcate from the middle of each side to the corners of the AU, because this reduces the chance of error of the 90º angles of the corners by half, compared to demarcating the angles from the corners of the unit.

In irregular spacing, you can measure the sides to get the actual area of the AU. After demarcation, the AUs are measured as follows (Figura 5):

- An X is drawn between the internal trees of the AU vertices and the three external trees of the vertices of the polygon formed by the AU border lines;
- Measure the 4 sides of the AU (A, B, C and D) from the center of each X formed at the vertices;
- The area (a) of the sampling unit in the Figura 5 is calculated by:

$$a = [(A+C)/2] * [(B+D)/2]$$

UA = 20m x 30m = 600 m²
Espaçamento regular de 2m x 3m

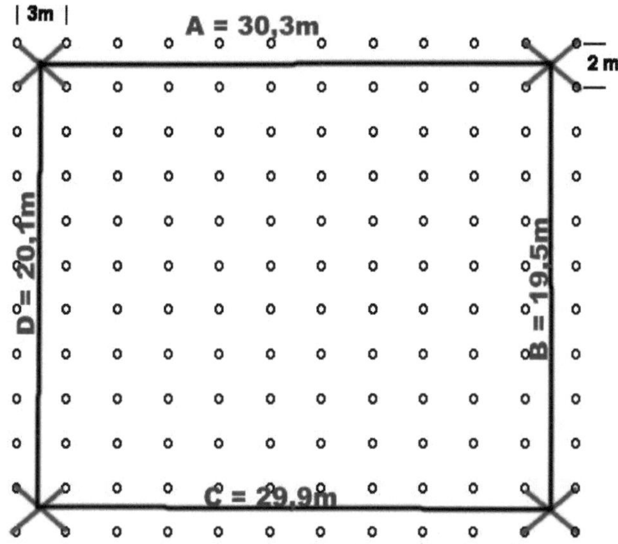

ÁREA DA UA = [(30,3+29,9)/2] x [(20,1+19,5)/2]
ÁREA DA UA = 595,98 m²

FIGURA 5 - Measuring the actual area of a square or rectangular sampling unit.

2.8.3 Demarcation of circular sampling unit

- The size of the radius is defined and the AU is demarcated;
- A tree whose base is on the boundary line with more than half of it inside the plot is counted;
- If the base is exactly halfway down the plot, the even trees are included; in other words, the 1st is not included, the 2nd is included, the 3rd is not included and so on.

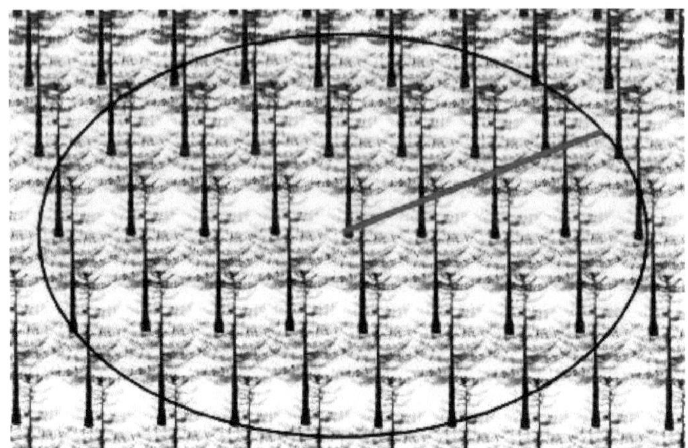

FIGURA 6 - Circular UA demarcation.

2.8.4 Demarcation of the Bitterlich sampling unit

The demarcation starts from the location of the center of the unit and proceeds as follows:

- The center of the AU is demarcated;
- From the center, trees with a DBH wider than the chosen angle are counted, preferably always starting from the north of the unit and going clockwise;
- Doubtful trees such as No. 4 (Figura 7) should have their doubts resolved by the equation:

$$d = R / [\ 50 \sqrt{(1/K)}\ ]$$

Where: d=limit diameter in m; R=distance from the center of the AU to the tree in m; K=angle numbering factor.

- If the calculated d limit is equal to or greater than the measured d of the tree, the tree is excluded.
- Trees 1 and 8 in this example are outside the AU;
- Tree 4 is doubtful, with d=35cm at a distance of 10m, and is determined to be within the AU, as the

calculated limit d [d = 10 / [50 √(1/3) = 34.64 cm] is less than the tree's diameter;

- We counted 7 trees in the AU of this example, resulting in G = 7x3 = 21 m²/ha.

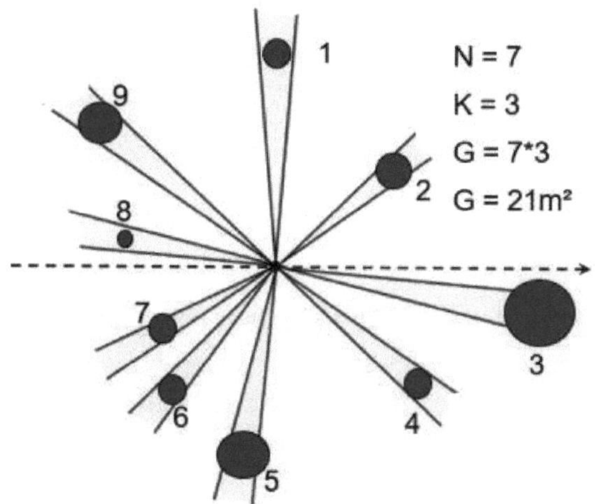

FIGURA 7 - Bitterlich sampling unit.

2.8.5 Demarcation of quadrant sampling units

- Mark the survey lines and the sample points on the lines; from the sample points, form right angles to both sides of the survey line, forming a cross (Figura 8);
- The tree closest to the sample point in each quadrant is measured;
- The sampling unit is made up of the four trees measured in the quadrants.

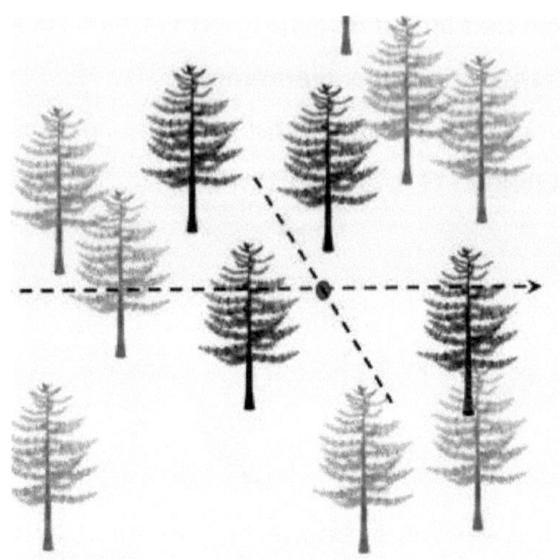

FIGURA 8 - Quadrant sampling unit.

2.8.6 Demarcation of the Prodan sampling unit

- Demarcate the survey line and the sample points on it;
- From the center, the six closest trees are sampled.

FIGURA 9 - Sample unit of the 6 Prodan trees

2.9 Trees bordering the sampling units

A border tree is one whose trunk base is on the boundary line of the sampling unit; if it is more than halfway inside the plot, it is included in the plot, otherwise it is disregarded. However, if there is any doubt as to whether it is inside or outside the sampling unit, i.e. if the base of the tree is exactly half inside the plot and half outside, the doubtful border trees are ordered from "1" to "n" and only those with an even order number are included in the plot; i.e. the 1st is not included, the 2nd is included, the 3rd is not, and so on.

2.10 Types of surveys in terms of the percentage of trees measured in the population

There are two ways of obtaining data on a forest population or community:

- Census - Measuring all individuals;
    - It is the complete enumeration of all the individuals in the population;
    - It is rarely justified in production forests, except in special cases such as forest management in the Amazon rainforest;
    - It is used in afforestation, urban forests and arboretums;
    - The results obtained are the population parameters;
    - The results are called population parameters and are represented by Greek letters.
- SAMPLING - Measuring a fraction of individuals.

- It is the collection of data from a fraction of the individuals in a population, by means of statistically distributed sampling units that are significantly smaller than the area of the population, with a certain probability of error associated with the results obtained;
- It is used in conventional forest inventories;
    - The results obtained are population statistics that estimate the parameters of the population;
    - They are represented by lowercase Arabic letters for individual tree averages;
    - The averages per hectare are represented by capital letters of the Arabic alphabet.

2.11 Required Precision and Sample Error Limit

The precision required in sampling depends on the following characteristics:

- Purpose of sampling;
- Financial resources available;
- Criticality:
    - Legislation - imposes restrictions and requirements;
    - Risks arising from imprecision.

Precision refers to the probability of confidence in the mean (P). It is common to use a confidence probability for the mean of

95% in inventories of planted forests and 90% in inventories of natural forests. The probability of error (p) is calculated as:

$$p = \pm (1 - P)$$

Where: p = probability of error; P = probability of confidence for the mean.

As an example: if we want a probability of confidence (P) for the mean of 95%, the probability of error (p) will be:

$$p = \pm (100\% - 95\%) = \pm 5\%$$

The sampling error is due to the part of the forest area that was not measured, represented by the difference between the sample mean and the population mean.

The maximum relative sampling error limit (LE%) is set before sampling. It is the percentage of error allowed in relation to the mean. The Limit of Error is the value represented by the amplitude between the lower and upper limits of the mean. It is calculated as twice the modulus of the probability of error, according to the equation:

$$LE\% = 2\ |\pm p|$$

Where: LE% = admitted sampling error limit; p = probability of error (p=1-P).

Example: If we want a 95% confidence level for the mean and the probability of error is $\pm 5\%$, the Error Limit will be calculated as:

$$LE\% = 2\ |\pm 5\%| = 10\%$$

The absolute sampling error limit (LE) is found by multiplying the relative sampling error limit (LE%) by the sampling average of the pilot inventory, as follows:

$$LE = LE\% \cdot \bar{y}$$

Where: LE = Absolute sampling error limit; LE% = Relative sampling error limit; $\bar{y}$ = average calculated in the pilot inventory.

Example: if the sample average ($\bar{y}$) in the pilot inventory is 250 m³/ha, and the relative sampling error limit (LE%) is 10%, the absolute sampling error limit (LE) will be:

$$LE = 10\% \,.\, 250\ m^3/ha = 25\ m^3/ha$$

After the field survey and statistical calculations of the final inventory, the sampling error that actually occurred in the sampling is also calculated to check whether it was below the accepted Error Limit. When the sampling error is greater than the allowed error, it may be necessary to intensify the sampling to bring it within the desired standards.

2.12 Precision and Accuracy

Precision refers to sampling error (easier to obtain) - lack of precision is represented by the deviation of the sample from the estimated mean, or standard error of the mean.

Accuracy refers to the deviation of the sample from the population parameter including measurement errors and other non-sampling errors, which is difficult to obtain.

2.13 Measurement errors

The most common measurement errors relate to:

- Forest area;
- Plot areas;
- Doubtful trees;
- Precision and instrumental calibration;

- Measuring point;
- Visual acuity;
- Operator attention.

2.14 Limit of sampling error depending on the objective

The error limit also depends on the type of forest, as well as the objective and size of the forest area to be inventoried.

With regard to the purpose of the inventory, the error limit is generally established as:

- Pre-cut timber inventory:
  - Native forest (Amazon): census of 100% of trees with d>=50cm, i.e. there is no sampling error, as all trees with a diameter of 50cm or more are measured; for smaller trees that are sampled, the LE% can be between 5% and 10%;
  - Planted forest: in pre-cut inventories, the error limit is set at between 2.5% and 5%; in continuous forest inventories, it is generally 10%.
- Phytosociological or ecological inventory:
  - The LE% is between 10% and 20% of the basal area per ha;
  - The stabilization of the species curve by the sampling area should be taken into account (the number of species increases as the sampling area increases until it stabilizes).
- Continuous forest inventory for growth studies:

- In this case, the most important thing is that the individual measurements are accurate and that the same tree with the same numbering is measured on each occasion;
- a sampling error limit (LE%) is normally accepted for each occasion:
  - up to 10% in planted forests;
  - up to 20% in natural forests.

2.15 Error due to profit reduction

An example of determining the error limit as a function of the loss in profits, independent of the cost of sampling, is given below.

Consider a timber industry that sells its products at R$1000.00/m³ with a profit of R$100.00 (10%). It uses 2 m³ of standing timber for each m³ of product manufactured;

Standing timber costs R$25.00/m³. Therefore, producing 1% more standing timber without increasing sales (1% . R$25.00 = R$0.25) implies a loss of (2m³ . R$0.25), representing R$0.50 of profit per m³ sold, i.e. 100 . R$0.50 / R$100, implying a loss of 0.5% of profit per m³ sold.

If the industry decides not to commit more than 5% of its profit to guaranteeing the wood supply for the mill, what should the sampling error limit (LE%) be?

Allowed loss = R$100.00/m³ . 5% = R$5.00 / m³

*LE% = 100 * ((R$5.00/m³) / (R$25.00*2m³))*

*LE% = 100 * 5 / 50 = 10%*

2.16 Sampling Intensity

It is the proportion of the total area of the sample in relation to the area of the population, calculated by:

$$IA = 100 * AA / AP$$

Where: IA = sampling intensity in percentage; AA = total sampled area (ha); AP = total population area (ha).

It is also interpreted as the sampling fraction:

$$f = n / N$$

Where: f = sampling fraction; n = number of sample units measured; N = potential number of sample units from the entire population.

The sampling intensity can be established according to:

- Mean ($\bar{Y}$) and variance (s^2) and maximum allowed error limit (LE%);
- Sampling costs;
- Profit reduction.

2.16.1 Sample Intensity as a function of mean ($\bar{y}$), variance (S^2) and error limit (LE%)

- finite populations

$$n = (N. t^2 . s^2) / [(N . LE^2) + (t^2 . s^2)]$$

- infinite populations

$$n = t^2 . s^2 / LE^2$$

Where: n = sample size or total number of units to be sampled; t = Student's t-distribution value, with error probability p and n-1 degrees of freedom; s^2 = variance; LE^2 = square of the admissible sampling error limit; N = total potential number of units in the population; n = number of sampling units measured in the pilot inventory.

2.16.2 Sampling intensity according to costs and available resources

The sampling intensity as a function of inventory costs is calculated by:

$$n = (Ct - C0) / C1$$

Where: n = number of sampling units; Ct = total cost of the inventory; C0 = fixed costs of planning, equipment, analysis and drawing up the inventory report; C1 = average cost per sampling unit (displacement + demarcation + measurement).

3. SAMPLING METHODS

Sampling methods are the ways of approaching a single sampling unit. Sampling methods are related to the type of sampling unit used. The main ones are classified into the 4 types listed below:

- Fixed area sampling units
  - Square
  - rectangular
  - circular
- Point sampling units
  - Nearest tree (common for cubing)
  - Bitterlich
  - Prodan (6 trees)
  - Quadrants
- Line or strip sampling units
  - Nearest neighbor
  - Transects (linear and strip)
  - Strand
- Empirical
  - 3-P
  - visual

3.1 Fixed area sampling unit method

3.1.1 Description of the fixed area method

In this method, the sampling units have a regular and fixed area for all units.

The units can have any regular shape such as square, circular or rectangular, but all units have the same shape and size.

It is the preferred method and considered the most accurate.

The adjustment of the unit size should be such that it encompasses most of the local forest variability in order to reduce the variability between the units and, consequently, resulting in a smaller number of units for the maximum desired error.

To transform the values for the frequency of trees in the sampling unit (n), average individual basal area (g) and average individual volume (v) into average values per hectare (N, G and V), the proportionality factor can be used. The proportionality factor is the ratio between the area of 1 ha and the area of the sampling unit, and is calculated by:

$$F = A / a$$

Where: A = 10000 m² (1 ha); a = area of the sampling unit in m².

Using the proportionality factor, the values of N, G and V are calculated by:

- Number of individuals, or frequency per hectare (N)

$$N = n \,.\, F = n \,.\, A / a = n \,.\, 10000 / a$$

Where: N = number of individuals per hectare; n = number of individuals found in the sampling unit; A = 10000 m² (1 ha); F = area proportionality factor; a = area of the sampling unit in m².

- Basal area per hectare (G)

$$G = \left(\sum_{i=1}^{n} g_i \right) . F$$

Where: G = basal area in m² per hectare; gi = individual basal area of tree i in m² ($g_i = \pi . di^2/4$); di = diameter at breast height of tree i; n = number of trees in the sampling unit; F = area proportionality factor.

➢ Volume per hectare (V)

$$V = \left(\sum_{i=1}^{n} v_i \right) . F$$

Where: V = volume in m³ per hectare; v_i = volume of tree i in m³; n = number of trees in sampling unit i; F = area proportionality factor.

3.1.2 Advantages of the fixed area method

The main advantages of using fixed-area plots are listed below:

- ➢ All statistics are obtained directly from each sampling unit;
- ➢ It's practical and simple;
- ➢ It is the most widely used in forest inventories;
- ➢ There is a high correlation on multiple occasions.

3.1.3 Square or rectangular sampling units

In forest plantations, care should be taken with square or rectangular plots, due to the irregular spacing between trees. The spacing between rows is usually demarcated by furrowing the area, resulting in rows that are not perfectly straight and equidistant. And the trees are rarely aligned in both directions, with the trees not having a corresponding position between one row and the next. Some professionals claim that these differences in area from one plot to another compensate for each other, i.e. one difference more compensates for one difference less in relation to the sizes of the many sampling units in a stand. However, it has been observed that the differences tend to be systematic, either upwards or downwards, depending on the topography and the operator of the furrowing machine or the digger, which leads to sometimes serious errors when calculating the averages per hectare.

Due to irregularities in the spacing between trees in plantations, many have preferred circular plots to square or rectangular ones. Others opt for square or rectangular plots, but correct the area of each plot individually, which is more appropriate than using a fixed value and gives more accurate results per hectare than without correcting the areas of the plots individually. The marking of the units and the correction can be done as in Figura 10described below:

- Using a cross, or a tape measure, measure a right-angled triangle with sides of 3, 4 and 5 m to mark two orthogonal lines from the center of the plot, one of the lines in the direction of the planting line;
- Then mark the points on the edges of the plot, halfway down the sides of the plot, where each line starts from the center;
- From the four edge points in the middle of the unit, the edge lines are extended to the four corners of the plot, creating the boundaries of the plot area;
- The marginal trees of the plots are marked with paint and an x at the entrance to the plot;
- The measurement is taken from the left corner (1), going back and forth along each line;

- In continuous inventories involving individual growth, it is necessary to number all the trees in the plots, so that the number of the measured tree is always the same on every measurement occasion;
- Trees on the boundary: the first tree found on the boundary of the plot from the corner where the measurement begins is not included in the plot, the second is included, the third is not included, the fourth is included, and so on;
- In the 4 corners of the plot, we try to locate the 4 original tree planting sites near each of the corners and place a stake at the intersection of the two diagonals formed by the four trees;
- The actual distances are measured (values in brackets in Figura 10) from one stake to another at the corners of the plot, corresponding to sides A, B, C and D;
- The actual area of the plot is calculated using the equation:

$$\text{Area of the plot} = (A+C) \,.\, (B+D) / 4$$

A Figura 10 shows the marking of a plot with 20m sides, with sides A=20.35m, B=19.90m, C=20.35m and D=19.64m measured, resulting in a real area of 402.32m².

3.1.4 Circular sampling units

In the circular plots (Figura 11) the dimensions of the AU are not influenced by the spacing of the planting and irregular spacing does not influence the size of the plot, it is enough to measure the radius correctly horizontally.

It is good practice to mark the first and last trees in each planting line to measure the trees, as this makes it easier to see during the measurement; in the measurement sequence there are shorter lines at the edge and longer ones in the middle, and it is difficult to identify the trees at the beginning and end of each line without marking them.

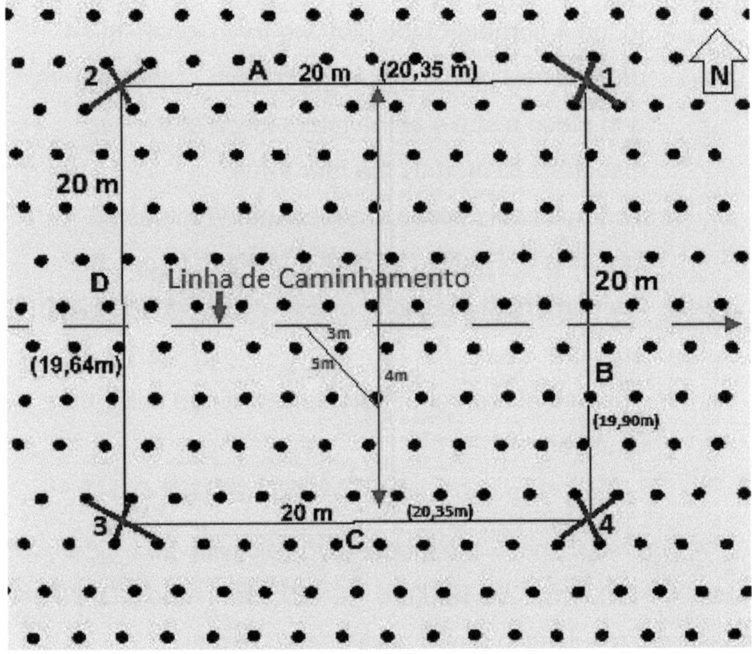

FIGURA 10 - Marking and correcting the area of rectangular or square sampling units

Circular sampling units have been preferred because they are less prone to error in terms of plot size. Care must be taken to maintain horizontality when measuring the radius to demarcate the plot.

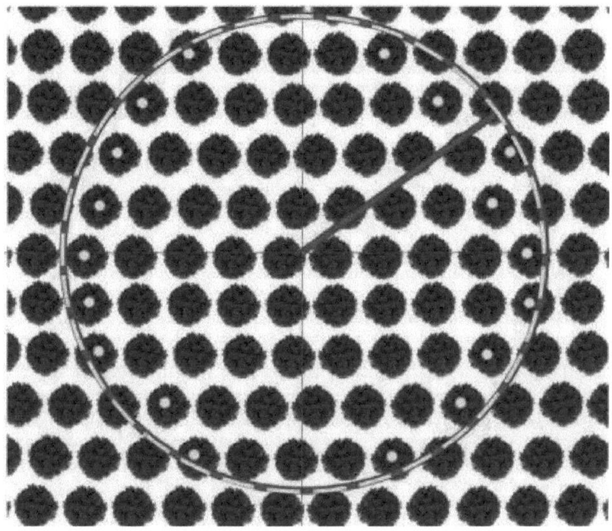

FIGURA 11 - Circular plot.

In Tabela 2 a comparison is made of the advantages and disadvantages between square or rectangular plots and circular plots.

TABELA 2 - Comparison between square and circular plots.

Square or Rectangular Plots	Circular plots
• The size of the plots is influenced by the uneven spacing of the planting; • Uneven spacing means measuring the sides to correct the dimensions of the plots; • The sequence in which the trees are measured follows lines of the same length, making it easier to orientate them in the plot and faster and easier to measure them.	• Plot dimensions are not influenced by planting spacing; • Uneven spacing does not influence the size of the plot, just measure the radius correctly; • The measurement sequence has shorter lines at the edge and longer ones in the center, making it more difficult to orient yourself on the plot during the measurement.

3.2 Variable area unit methods

3.2.1 Nearest Tree Method

3.2.1.1 Characteristics of the method

In this method, sampling points are distributed systematically or randomly over the vegetation under study and the tree closest to the sampling point is found.

The sampling unit in this method consists exclusively of the tree closest to the sampling point.

The area of the sampling unit (a) is calculated by squaring the corrected distance (D) from the tree to the sampling point, as follows:

$$a = (2 \cdot D)^2$$

Where: a = area of the sampling unit; D = distance from the nearest tree to the sampling point.

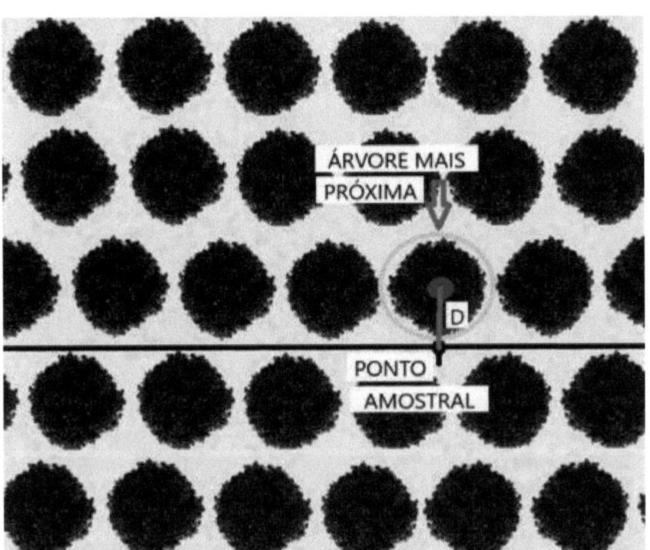

FIGURA 12 - Installation of the sampling unit using the nearest tree method.

The area proportionality factor (F) used to convert the statistics per plot into estimates per hectare is calculated by:

$$F = A / a$$

Where: F = proportionality factor; A = 10000 m² (1 ha); a = area of the sampling unit in m².

The other calculations are carried out as in the fixed area sampling method.

3.2.1.2 Advantages, disadvantages and precautions

The advantages of the method include:

- Easy to apply;
- No special tools are needed to demarcate the units;
- It can be applied to any type of vegetation;
- Allows the selection of a single individual per unit when selecting trees for cubing.

The main disadvantage is the need for a large number of sampling units to represent the population, due to the excessive variation between them.

The distance from the sampling point to the center of the trunk at the base of the tree must be correctly measured in order to get the correct area of the sampling unit.

3.2.1.3 Nearest tree method in cubing sampling

When sampling for accurate tree cubing, the nearest tree method can be used. There is no definitive rule for determining the number of trees needed for cubing, but it should be such that it allows for residue analysis. The ideal would be to have 30 individuals per diameter class in order to set up a residue distribution curve per class for the analysis, but the costs would be very high. Some researchers admit that from a total of 20 individuals it would be possible to adjust a valid regression equation, but sometimes it is not possible to reach this number. So the biggest limitation is the cost of obtaining the cubing data. However, the quality of the results of the analysis of variance of the regression and its validation by residue analysis will indicate whether the number of individuals cubed was sufficient.

In order to sample the trees to be felled, it is necessary to know the diameter data of the trees measured in the plots. With the tree diameter data collected from the inventory plots, the smallest (LI_{DAP}) and largest diameters (LS_{DAP}) are determined. The number of diameter classes for the population is then calculated using Sturges' rule for samples containing up to 2000 individuals, or Floriano's rule for large samples (>2000 individuals), or any other rule the researcher prefers. The number of trees to be sampled must be the same for each diameter class of trees to allow for the analysis of residues graphically or by statistical methods.

3.2.1.4 Calculating the number of diameter classes

- Sturges: $k = 1 + 3.3 \cdot \log N$

  Where: k = number of diameter classes; N = total number of trees listed in the sample; log = base 10 logarithm; Example: consider a sample containing 1500 trees, k = 1 + 3.3 * Log 1500 = ~12 DBH classes;

- Floriano: $k = N^{0,175} \cdot \Delta d^{0,3}$

  Where: k = number of diameter classes; N = total number of trees listed in the sample; k = number of diameter classes; N = total number of trees listed in the sample; Δd = range of diameters in the sample ($=LS - LI_{DAPDAP}$); example: sample containing 8000 trees, with the largest diameter of 45 cm and the smallest of 10 cm; Δd =

45cm-10cm = 35cm; k = $(8000)^{0,175} . (35)^{0,3}$ = 4.82 . 2.91 ≈ 14 DBH classes.

An example is shown in Figura 13In the second case, the sample has 8000 trees with the smallest diameter of 10 cm and the largest of 45 cm, and by calculating the number of classes, 14 classes have been obtained, and it was planned to cover 70 trees, so 70 trees / 14 classes = 5 trees per DBH class must be sampled. Consequently, 70 sampling points should be distributed over the forest area and at each sampling point a DBH class is drawn, with 5 trees being randomly selected for each DBH class, i.e. each sampling point will receive a value from 1 to 14, representing the DBH class of the tree to be sampled at the point. The tree to be sampled at each sampling point will be the one closest to the point with the diameter belonging to the class drawn.

3.2.2 Bitterlich method

Walter Bitterlich was a renowned forest scientist (1908-2008) who, in 1947, proposed the angular numbering sampling method.

The statement of the Bitterlich method is as follows: "The number of trees (n) in a stand whose diameters at a height of 1.3 m (d) seen from a fixed point appear larger at a given angle (α) is proportional to their basal area per hectare (G)".

O retângulo representa a área da floresta, com a sistematização de 70 pontos amostrais localizados nos cruzamentos das linhas verticais e horizontais.

Classes de DAP (cm)			
Nº	>= LI	< LS	CC
1	10,00	12,50	11,25
2	12,50	15,00	13,75
3	15,00	17,50	16,25
4	17,50	20,00	18,75
5	20,00	22,50	21,25
6	22,50	25,00	23,75
7	25,00	27,50	26,25
8	27,50	30,00	28,75
9	30,00	32,50	31,25
10	32,50	35,00	33,75
11	35,00	37,50	36,25
12	37,50	40,00	38,75
13	40,00	42,50	41,25
14	42,50	45,00	43,75

14	10	12	10	9	8	12	14	3	4
1	3	11	7	6	12	2	14	3	9
12	1	1	9	13	14	13	13	9	2
8	5	4	6	6	2	10	10	4	1
5	2	11	9	2	12	7	6	8	5
7	11	11	10	11	14	4	5	4	3
8	7	13	14	13	3	6	1	8	7

Sendo os pontos numerados da direita para esquerda e de cima para baixo, no primeiro ponto amostral deve ser amostrada a árvore mais próxima da classe 14 com DAP de 42,50 a 45,00 cm, no segundo da classe 10 com DAP de 32,50 a 35,00 cm e assim por diante.

FIGURA 13 - Systematic distribution of 70 sampling points over the forest area to select trees for rigorous cubing.

3.2.2.1 *Method demonstration*

Bitterlich developed the method from a bar with an eyepiece 1 meter away to an objective lens 2 centimeters wide (Figura 14). The trees are therefore selected with a probability proportional to the diameter of the trees.

With the Bitterlich bar, you count the trees with a diameter wider than the aperture of the lens from a point fixed in a 360° rotation. It is suggested that you always start in the north and count clockwise. The number of trees counted (n) is equal to the basal area per hectare (G):

$$G = n$$

Where: G = Basal area per hectare (m^2/ha); n = number of trees counted in the sampling unit.

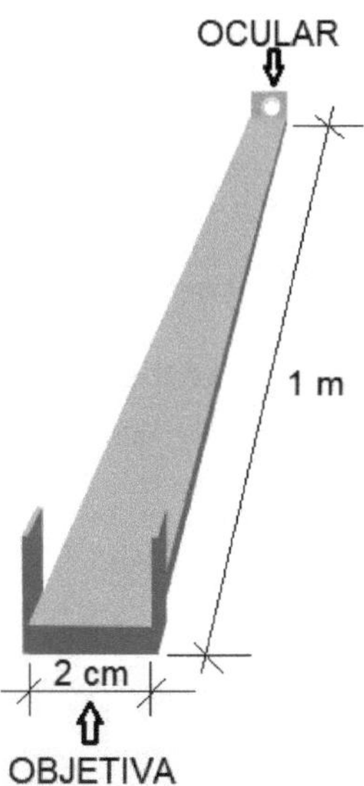

FIGURA 14 - Bitterlich bar for angle count sampling.

Considering the Figura 15 using the Bitterlich bar, the basal area per hectare is calculated for one tree using the equation:

$$G = K$$

Where: G = Basal area per hectare (m²/ha); K = Basal area factor, also called angular numbering factor.

Since G = n and G = K, G = n . K, then the K value of the Bitterlich bar with an objective aperture of 2% is equal to 1.

The basal area for more than one counted tree is determined by:

$$G = K \, . \, n$$

Where: G = Basal area/ha; K = Basal area factor; n = Number of individuals counted at the sampling point with diameters greater than the opening angle (α).

In Figura 15consider the relationship between the measurements:

$$a \, / \, L = d \, / \, R$$

Where: L = length of the bar; a = angular opening; d = diameter at 1.3 m above the ground; R = radius of the plot.

If the relationship is true:

$$a/L=d/R_{ii}$$

E, being the surface area of the plot (S):

$$S = \pi \, . \, R^2$$

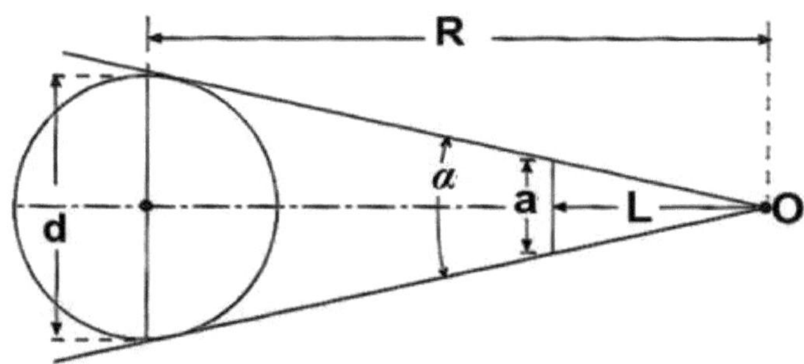

FIGURA 15 - Variables used to develop the Bitterlich method: α = angle for angle counting; a = width of the Bitterlich bar lens corresponding to the angle α; L = distance from the eyepiece to the bar lens; O = eyepiece; d = tree diameter at 1.3 m height; R = distance from the bar eyepiece to the center of the tree trunk.

The basal area of a tree (g_i) is calculated as:

$$g_i = \pi \, . \, d_i^{\,2} / 4$$

Where: g_i = basal area of tree i; a = lens width (m); L = bar length (m); R_i = distance from the sampling point to the center of the trunk of tree i (m); S = plot area (m^2); d_i = diameter of tree i (m).

Therefore, the basal area (g_{UA}) of the sampling unit corresponding to tree "i" is given by:

$$g_{UAi} = g_i / S = (\pi . d_i^{\,2}/4) / \pi . R^2$$

$$g_{UAi} = (d_i^{\,2}/4) / R^2 = (1/4).(d_i^{\,2}/R^2)$$

$$g_{UAi} = (1/4).(d_i /R)^2$$

For 1 hectare (10000m²), multiply g_{UA} by 10000m²:

$$G_{UAi} = 10000 \,.\, g_{UAi}\text{, or}$$

$$G_{UAi} = 10^4 \,.\, (1/4).(d_i /R)^2$$

The slope of the terrain means that the horizontal distance to the diameter of the trees at a height of 1.3 m has to be calculated, as the distance is the side of a right triangle and the distance to the tree is the hypotenuse of this triangle.

Based on this principle, in 1955, Bitterlich developed the apparatus known as the Relascope (Figura 16).

Funcionalidades:

- Amostragem de Bitterlich;
- Medição de altura;
- Medição de diâmetros a diferentes alturas.

- É baseado no princípio de Bitterlich;
- Possui bandas correspondentes a diferentes ângulos de abertura.

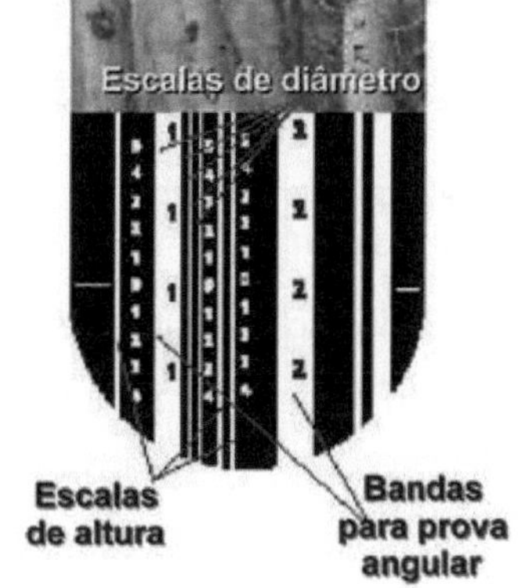

FIGURA 16 - Mirror relascope.

Relascope automatically corrects the slope in the line of sight, making it possible to find the correct basal area of the stand in square meters per hectare regardless of the slope of the terrain.

Currently, there is a choice of electronic devices with the same functionality as the Relascope, such as the Criterion dendrometer (Figura 17).

FIGURA 17 - Criterion dendrometer.

3.2.2.2 Angle Numbering Factor (K), or instrumental constant

Considering the occurrence of a single tree larger than the angle (α) in the sampling unit (N=1), and making:

$$K = GUA_i$$

We obtain the value of the angle numbering factor (K), which is given by:

$$K = 10000.\ (1/4).(d\ /R_{ii}\ )^2$$

The basal area per hectare (G) is calculated as the number of trees per hectare (N) multiplied by the average basal area of the trees ($\bar{g}$) of the sampling unit, as:

$$G = N\ .\ \bar{g}$$

Continuing analogously:

$$G = N \,.\, GUA_i$$

By substituting GUAi, the Bitterlich equation is obtained:

$$G = N \,.\, K$$

It is understood that there is a critical radius (R_i) for each tree at the sampling point to be counted, as it has a diameter (d_i) greater than the opening of the angle corresponding to a certain numbering factor (K), calculated by:

$$K = 10000.\ (1/4).(d\ /R_{ii}\ )^2$$

Or:

$$K = 2500\ (d_i\ /R_i\ )^2,\ R_i = d_i\ \sqrt{(2500/K)},\ d_i = R_i\ /\ \sqrt{(2500/K)}$$

3.2.2.3 Doubtful trees

Doubtful trees have been treated in three different ways in angle count sampling:

1. By measuring the horizontal distance from the center of the AU to the center of the tree trunk at 1.3 m height, the limit diameter is calculated by $d_{lim} = R/\sqrt{(2500/K)}$; if d_{lim} is smaller than the tree's DBH it is included in the AU; this is the most correct way (Tabela 3);
2. Sorting the doubtful trees and excluding the 1st, including the 2nd, and so on;
3. Counting doubtful trees as 0.5 instead of 1 to calculate G=N.K; this is the worst method and complicates the calculation of averages, as the average has to be calculated using half the diameter or height of the trees.

TABELA 3 - Examples of trees in a sampling unit that should or should not be counted depending on the band (numbering factor), the distance to the center of the sampling unit and the diameter (d) of the tree.

Tree	K	d (cm)	Distance to the tree (m)	Critical radius (m)	Count? (yes/no)
1	1	28	7,0	14,0	Yes
2	2	28	7,0	7,0	No
3	3	28	7,0	4,7	No
4	4	28	7,0	3,5	No
5	1	35	8,0	17,5	Yes
6	2	35	8,0	8,8	Yes
7	3	35	8,0	5,8	No
8	4	35	8,0	4,4	No

3.2.2.4 Sampling by angular numbering test

First, the center of the AU is marked. Starting in a clockwise direction from the north, count the trees with a DBH wider than the chosen angle in a 360° turn. Doubtful trees such as No. 4 should have their doubts resolved by the equation:

$$d_{LIM} = R / [\, 50 \sqrt{(1/K)}\,]$$

Where: dLIM=limit diameter in m; R=distance from the center of the AU to the tree in m; K=angle numbering factor.

If the calculated d_{Lim} is equal to or greater than the tree's measured d, the tree is excluded.

In Figura 18 trees 1 and 8 are outside the AU; tree 4 is doubtful, with d=35cm at 10 m distance it is inside because the diameter of the tree is greater than the limit diameter (d_{LIM}) calculated by:

$$d = R_{LIM} /[50\sqrt{(1/K)}] = 10/[50\sqrt{(1/3)}] = 34.6cm$$

In this AU, 7 trees were included and counted with a numbering factor of K=3, resulting in 21 m²/ha of basal area.

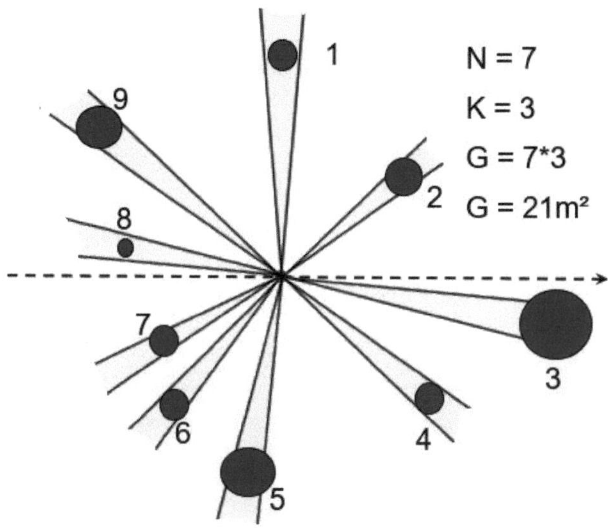

FIGURA 18 - Bitterlich sampling.

3.2.2.5 Determining the average of the variables in the sampling unit

3.2.2.5.1 Average basal area per hectare

The basal area per hectare (G) in m²/ha is obtained directly by multiplying the numbering factor (K) by the number of trees (n) included in the sampling unit:

$$G = K \, . \, n$$

3.2.2.5.2 Arithmetic mean diameter ($\bar{d}$)

It is calculated using the diameters weighted by the basal area of the trees, using the equation:

$$\bar{d} = \frac{\sum_{i=1}^{n} \frac{d_i}{g_i}}{\sum_{i=1}^{n} \frac{1}{g_i}}$$

Where: $\bar{d}$ = average diameter of the sampling unit; n = number of trees selected at the sampling point; di = individual diameter of tree "i"; g_i = basal area of tree "i" (m^2).

3.2.2.5.3 Arithmetic mean height ($\bar{h}$)

It is calculated using the heights weighted by the basal area of the trees, using the equation:

$$\bar{h} = \frac{\sum_{i=1}^{n} \frac{d_i}{g_i}}{\sum_{i=1}^{n} \frac{1}{g_i}}$$

Where: $\bar{h}$ = average height of the sampling unit; n = number of trees selected at the sampling point; hi = individual height of tree "i"; g_i = basal area of tree "i" (m^2).

3.2.2.5.4 Frequency per hectare (N)

The Frequency (N), or number of trees per hectare, is obtained from the equation:

$$N = K.\sum_{i=1}^{n} \frac{1}{g_i}$$

Where: N = frequency of trees per hectare; K = basal area factor; gi = basal area of tree "i" (m^2).

3.2.2.5.5 Volume per hectare (V)

The volume per hectare is determined using the equation:

$$V = K.\sum^{n} \frac{v_i}{g_i}$$

Where: V = Volume per hectare (m^3/ha); K = basal area factor; n = number of trees selected at the sampling point; v_i = volume of selected tree "i" (m^3); g_i = basal area of selected tree "i" (m^2).

3.2.2.6 Advantages of angle number sampling

- Obtain the basal area per hectare immediately;
- Quick volume calculation when the average form factor is known;
- Ease and speed of locating the sampling unit (point).

3.2.2.7 Disadvantages of sampling by angle numbering

- They generally require a greater number of units than fixed-area units;
- Higher cost than the fixed area method, in general, for the same accuracy;
- Special equipment required;
- Less precision in estimating volume;
- The trees measured may not be the same on two occasions in a continuous inventory, making them unsuitable for growth studies.

3.2.2.8 Care in sampling by angular numbering

- For greater precision, the distances from the center to the middle of the trunk at the base of the doubtful trees should be measured to determine whether they are inside or outside the sampling unit.

3.2.3 Prodan's 6-tree method

The Prodan method has some similarities to the Bitterlich method, except that it fixes the number of trees to be considered in the sampling unit.

The sampling unit is made up of the 6 trees closest to its central sampling point. Therefore, the area of the sampling unit is variable and selection is carried out with probability proportional to distance (PÉLLICO NETTO and BRENA, 1997).

The method has been used in studies of competition between trees.

3.2.3.1 Surface area of the sampling unit (S)

The surface area of the sampling unit is calculated as the area of a circle with a radius corresponding to distance from the 6th tree furthest from the center of the unit among the 6 trees at the sampling point , according to the equation:

$$S = \pi . R_6^2$$

Where: S = surface area of the sampling unit; R_6 = distance from the 6th tree furthest from the center of the unit among the 6 trees at the sampling point.

3.2.3.2 Frequency per hectare (N)

The number of trees per hectare, or frequency per hectare, is determined as follows:

$$N = \frac{55000}{\pi . R_6^2}$$

Where: N = frequency of trees per hectare; R_6 = distance from the 6th tree furthest from the center of the unit among the 6 trees at the sampling point.

3.2.3.3 Basal area per hectare (G)

The basal area per hectare is estimated by the equation:

$$\hat{G} = \frac{2500.\left[\sum_{i=1}^{n-1} d_i^2 + \left(\frac{d_6}{2}\right)^2\right]}{R_6^2}$$

Where: G = estimated basal area in m²/ha; n = number of trees counted in the sampling unit; di = diameter of tree of order i; d6 = diameter of tree of order 6; R6 = distance (radius) from tree of order 6 to the central point of the sampling unit.

3.2.3.4 Volume per hectare (V)

The volume per hectare is determined using the equation:

$$V = \frac{10000.\left[\sum_{i=1}^{n-1} v_i + \left(\frac{v_6}{2}\right)\right]}{\pi.R_6^2}$$

Where: V=volume in m³/ha; n = number of trees counted in the sampling unit; vi = volume of tree of order i; v6 = volume of tree of order 6; R6 = distance (radius) from the tree of order 6 (furthest) to the central point of the sampling unit.

3.2.3.5 Advantages of the Prodan method

- is practical and easy to apply;
- sampling is carried out quickly;
- there is no error in demarcating the sampling unit.

3.2.3.6 Disadvantages of the Prodan method

- bias of the estimators when the density of trees per hectare is very small or very large;
- It is not recommended for use in succession studies or forest management due to the small number of

trees per sampling unit, making it difficult to obtain good estimates for mortality and dominant height.

3.2.4 Quadrant method

This method was used by the American government when surveying the territory of the United States in the mid-19th century.

The sampling unit for the quadrant method is made up of two lines crossed orthogonally, forming 4 quadrants, with the tree closest to its center being included in each quadrant, for a total of 4 trees per sampling unit.

The coordinates of each tree are taken in relation to the two axes that start from the central point forming the quadrant, the distance to the tree located in the next quadrant, its diameter and other dendrometric variables of interest.

3.2.4.1 Area of the Quadrant sampling unit

The area of the sampling unit is defined as 4 times the average area per tree, while the average area per tree can be calculated as in Figura 19.

The equations used to determine the area per tree are:

$$a = x_i \,.\, y_{;i+1}$$

$$b = y_i \,.\, x_{;i+1}$$

$$Sum_a = a + a + a_{1234} + a = -44;$$

$$Sum_b = b + b + b_{1234} + b = +47;$$

$$Area\ per\ tree = |(Sum_a - Sum_b)/2| = 45.5\ m^2.$$

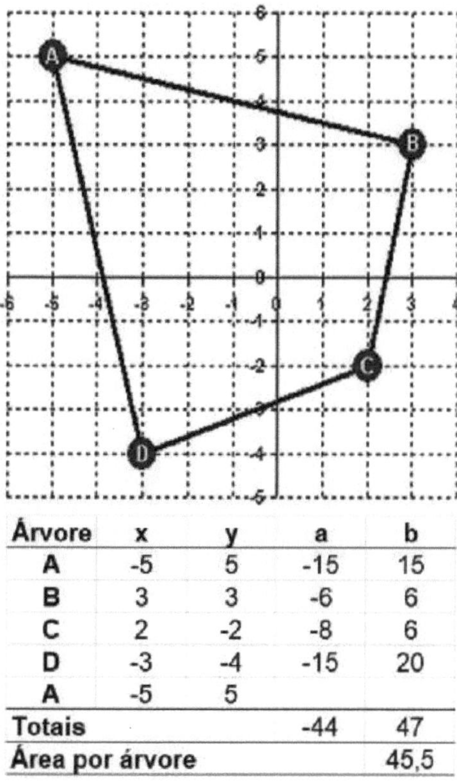

Árvore	x	y	a	b
A	-5	5	-15	15
B	3	3	-6	6
C	2	-2	-8	6
D	-3	-4	-15	20
A	-5	5		
Totais			-44	47
Área por árvore				45,5

FIGURA 19 -Determining the area per tree using the Gauss topographic method.

The number of trees per hectare is determined as follows:

- Average distance between trees = $\sqrt{45.5}$;
- Plot area = 4 x area per tree = 182 m²;
- N = 10000 . 4 / plot area = 219.78 trees/ha.

Considering the Figura 19the area per tree is 45.5 m² and the average spacing between trees is 6.75 m. The area of the quadrant will therefore be 182 m² (= 4 trees . 45.5 m²).

3.2.4.2 Area Proportionality Factor (F)

Proportionality factor used to convert unit statistics into estimates per ha:

$$F = A / a$$

Where: A = 10000 m² (1 ha); a = area of the sampling unit in m².

3.2.4.3 Frequency per hectare (N)

The number of individuals, or frequency per hectare, is calculated using the equation:

$$N = n \,.\, F = n \,.\, A / a = n \,.\, 10000 / a$$

Where: N = number of individuals per hectare; n = number of individuals found in the sampling unit; A = 10000 m² (1 ha); F = area proportionality factor; a = area of the sampling unit in m².

3.2.4.4 Basal area per hectare (G)

The basal area per hectare is found using the expression:

$$G = \left(\sum_{i=1}^{n} g_i\right).F$$

Where: G = basal area in m² per hectare; g_i = individual basal area of tree i in m² ($g_i = \pi.di^2/4$); d_i = diameter at breast height of tree i; n = number of trees in the sampling unit; F = area proportionality factor.

3.2.4.5 Volume per hectare (V)

The volume of the sampling unit per hectare (V) is estimated by the equation:

$$V = \left(\sum_{i=1}^{n} v_i\right).F$$

Where: V = volume in m³ per hectare; vi = volume of tree i in m³; n = number of trees in sample unit i; F = area proportionality factor.

3.2.4.6 Advantages

- Ease and speed in locating and delimiting sampling units.

3.2.4.7 Disadvantages

- Greater error in estimates due to the 90° exclusion angle;
- Need for a large number of UAs.

3.2.4.8 Care

- Correct measurement of the distance between trees, using the centers of the trunks at their bases as the start/end points of the measurement.

3.2.5 Strand method

The Strand sampling method can be considered a line method based on the Bitterlich method, with selection by probability proportional to diameter for sampling basal area per hectare (G) and by probability proportional to height for sampling volume per hectare (V).

The sampling unit is a line of length "L" on which all the trees that qualify for sampling are listed. The standard length of the Strand sampling unit is 15.7 meters (PELLICO NETTO E BRENA, 1997).

The survey is carried out in two stages:

- 1st) Sampling proportional to diameter using the Bitterlich method applied to lines;

- 2nd) Sampling proportional to height, where D ≤ (h/2), or h > tg 63°26'06".

The probability of inclusion (pi) of a tree in the sampling unit is given by:

$$p_i = R_i \,.\, L / A$$

Where: p_i = probability of including tree i; R_i = marginal radius of tree "i" included in the sampling unit; L = length of the sampling line; A = area of the stand.

Or, by the Bitterlich method:

$$p_i = d_i \,.\, L / K \,.\, A$$

Where: $K = d_i / R_i$ = basal area factor, or $R_i = d_i / K$; d_i = diameter of tree i; L = length of the sampling line; A = area of the stand.

Thus, the probability of a tree being included or not in the unit is given by the Bernoulli distribution (yes=1 / no=0).

3.2.5.1 Basal area per hectare (G)

According to Péllico Netto and Brena (1997), the unbiased estimator for basal area per hectare (G), for lines (sampling units) of any length, derived from the Bernoulli distribution and associated with the Bitterlich basal area factor (K), is as follows:

$$\hat{G} = \frac{\pi . \sqrt{K}}{2\,L} . \sum_{i=1}^{n} d_i$$

Where: $\hat{G}$ = estimated basal area in m²/ha; n = number of trees counted in the sampling unit; K= basal area factor; L = length of the sampling unit; d_i = diameter of the tree of order i; i = order number of the tree.

3.2.5.2 Frequency per hectare (N)

The number of trees per hectare, or frequency per hectare, is estimated using different equations to determine the basal area per hectare (G) and the volume per hectare (V), as follows:

3.2.5.3 Volume per hectare (V)

The volume per hectare can be obtained from an average shape factor (f) for the population, calculated by:

$$V = f\frac{1}{10}.\sum_{i=1}^{m} d_i^2$$

Where: V = volume in m³/ha; f = average form factor of the trees in the stand; m = number of trees counted in the sampling unit; di = diameter of the tree of order i; m = number of trees in the sampling unit; i = order number of the tree.

3.2.5.4 Advantages

- reducing the time needed for sampling;
- reduction of errors due to the definition of unit boundaries;
- the flexibility of the numbering factor means that an adequate number of trees can always be sampled;
- the survey can be carried out with low-cost devices;
- can be used for regeneration inventories.

3.2.5.5 Disadvantages

- difficulty in choosing trees when the understorey is dense;
- sampling errors occur when there are optical defects in the devices;

- is unfeasible as a unit for continuous surveys and growth estimates;
- basal area and volume per hectare are determined differently from each other;
- determining the volume implies determining the average form factor of the trees in advance.

3.2.5.6 Care

- doubtful trees should have their distance from the sample line measured for verification;
- a minimum distance must be established between the sampling lines so that there are no trees belonging to two sampling units at the same time.

3.2.6 Nearest neighbor method

In this method, the sampling unit consists of the tree closest to the sampling point and its nearest neighbor (Figura 20).

The distance between the tree closest to the sampling point and its nearest neighbor is multiplied by the correction factor 1.67 to determine the area of the sampling unit.

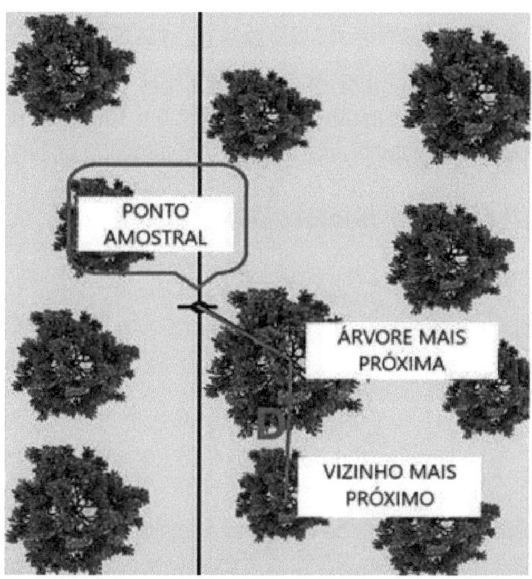

FIGURA 20 - Nearest neighbor method.

3.2.6.1 Area of the sampling unit (a)

The area of the sampling unit is calculated by squaring the corrected distance, as follows:

$$a = (1.67.D)^2$$

Where: a = area of the sampling unit; D = distance from the tree closest to the sampling point to the nearest neighbor; 1.67 = correction factor.

3.2.6.2 Area Proportionality Factor (F)

The area proportionality factor used to convert the statistics per plot into estimates per ha is calculated by:

$$F = A / a$$

Where: A = 10000 m^2 (1 ha); a = area of the sampling unit in m^2.

3.2.6.3 Frequency per hectare (N)

The number of individuals, or frequency per hectare, is calculated as follows:

$$N = n \,.\, F = n \,.\, A / a = n \,.\, 10000 / a$$

Where: N = number of individuals per hectare; n = number of individuals found in the sampling unit; A = 10000 m² (1 ha); F = area proportionality factor; a = area of the sampling unit in m².

3.2.6.4 Basal area per hectare (G)

The basal area per hectare (G) is found using the expression:

$$G = \left(\sum_{i=1}^{n} g_i\right).F$$

Where: G = basal area in m² per hectare; g_i = individual basal area of tree i in m² ($gi=\pi.d_i{}^2/4$); di = diameter at breast height of tree i; n = number of trees in the sampling unit; F = area proportionality factor.

3.2.6.5 Volume per hectare (V)

The volume of the sampling unit per hectare (V) is estimated by the equation:

$$V = \left(\sum_{i=1}^{n} v_i\right).F$$

Where: V = volume in m³ per hectare; vi = volume of tree i in m³; n = number of trees in sampling unit i; F = area proportionality factor.

3.2.6.6 Advantages

- Ease and speed of locating and demarcating AUs.

3.2.6.7 Disadvantages

- Error in the AU area due to the 180° exclusion angle;
- Large number of UAs required.

3.2.6.8 Care

- Correct measurement of the distance between the two trees, from the centers of their bases to the center of the AU.

3.2.7 Transect methods

The sampling unit is a line or strip (transect) cut across the study area, which can cross the entire area or have fixed lengths when studying forest profiles. It is mainly used in natural forest inventories for phytosociological studies, gradient studies and vegetation profiling.

To be able to analyze the variance of the population, it is necessary to use three transects or more and to be able to know the internal variance of the transects, it is necessary to divide them into parts of equal length, so that there are no less than three units per transect; for example: if there are three transects in the sampling with a length of 150, 350, 730 meters, they could be divided into subunits of 50 meters in length, leaving the first with 3 subunits, the second with 7 and the third with 14 complete subunits and one incomplete subunit of 30 meters in length.

3.2.7.1 Types of transects

- Linear transect
- Lane transect

3.2.7.1.1 Linear transect

Linear transects (Figura 21) are mainly used to make profiles at scales of 1:100 to 1:50 for trees and shrubs and 1:10 or larger for regeneration and herbs.

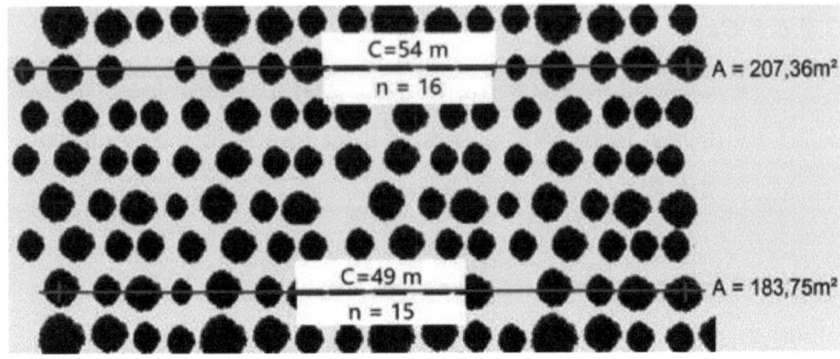

FIGURA 21 - Linear transects.

The trees on the sampling line are included in the transect. The length (L) of the sampling line is measured, from the center of the first trunk to the center of the last trunk in the line. The average distance (D) between trees on the line is calculated using the expression:

$$D = L / (n-1)$$

Where: D = average distance between trees in the sampling line; L = length of the sampling line; n = number of trees in the sampling unit.

The area of the sampling unit (a) is calculated by:

$$a = n * D^2.$$

The area proportionality factor (F) is calculated independently for each sample line by:

$$F = 10000\ m^2 / a$$

Where: F = area proportionality factor; a = area of the sampling unit in m²; 10000 m² = 1 ha.

The rest of the calculations are carried out as in sampling by fixed area units.

3.2.7.1.2 Strip transect

A belt or strip transect is an elongated plot of fixed width (Figura 22).

Transects cross the entire vegetation type under study; the strips can be subdivided into smaller plots to make it possible to calculate the variance within it. They are very useful for studies of natural vegetation in small forest areas up to a maximum of 500 hectares, with systematically distributed sampling units.

The formulas used for calculations are the same as those used for fixed-area sampling units, but all statistics must be transformed to the hectare, as transects can have different lengths, implying a different area for each one.

FIGURA 22 - Strip transects.

The width and length of the sampling unit should be such as to represent the population studied with maximum fidelity, and the transects should be divided into sub-units of fixed length to study the variability within them.

4. SAMPLING SYSTEMS

Sampling systems are characterized by the way the sample units are distributed and the statistical approach.

There are five basic types of distribution:

a) Random without restrictions:

In unrestricted random sampling, the units are randomly drawn from the entire population. In practice, the entire area of the population is divided into N potential units of the same size and n sample units are drawn to make up the sample. Another way is to define the size of the sampling units and the number needed and draw the location coordinates of each one over the population area.

b) Random with restrictions:

Some kind of restriction is imposed on the population before the sampling units are distributed, which can be strata or sampling levels. In stratum sampling, the population is divided according to the different characteristics it has, separating it according to the species that occur, different ages, etc. In multilevel sampling, the population is divided into first-level units, which are then redivided into smaller second-level units, and there may be more levels. Once the strata or levels have been formed, the sampling units are distributed randomly and independently in each level or stratum.

c) Systematic without restrictions:

The units are distributed systematically over the entire population. For lines or bands, an equidistance value is defined between lines or bands (D) and a distance value (d) is drawn within the range of the equidistance D, which will be the start of the first line or the middle of the first sample band. The other lines or strips will be distanced from each other by the equidistance value D. For grids, the total area is divided into equidistant vertical lines and equidistant horizontal lines and the coordinates of the first intersection of the lines are drawn by lot.

d) Systematic with restrictions:

The population is divided into strata or sampling levels. In stratum sampling, the population is divided according to the different characteristics it has, separating it according to the species that occur, different ages, etc. In multilevel sampling, the population is divided into first-level units, which are then redivided into smaller second-level units, and there may be more levels. Once the strata or levels have been formed, the sampling units are distributed systematically and independently in each level or stratum.

e) Mixed - part random and part systematic.

It is a restricted sample carried out on two or more levels. On at least one level the distribution is random and on at least one other level the distribution is systematic.

4.1 Unrestricted sampling statistics

4.1.1 Sampling intensity

Sampling Intensity (SI) is represented by the percentage of the forest area under study that was sampled. The AI is calculated by:

$$AI = 100 \,.\, n \,.\, a / A$$

Where: AI = Sampling Intensity in percentage; n = number of sampling units; a = area of the sampling unit in square meters; A = area of the population in square meters.

As an example, consider a forest plantation with an area of 100 hectares, or 1,000,000 m^2, subjected to a sample of 50 sampling units of 400 m^2 each, making up a sample of 50 x 400 m^2 = 20000 m^2, or 2 hectares; the sample therefore represents a Sampling Intensity (SI) of 2% of the area of the forest population, which makes it a finite population.

The sampling intensity can be calculated alternatively by the percentage of sampling units used to represent the population (n) and the potential number of sampling units (N). The potential number of sampling units (N) is calculated by dividing the total area of the population by the area of a single sampling unit. The sampling intensity

Populations whose sampling intensity is less than 2% are said to be infinite. When the sampling intensity is 2% or more, the population is said to be finite. Finite populations need to have their statistics corrected to reduce bias, using a multiplicative correction factor (CF) calculated by the expression:

$$FC = \left(\frac{N - n}{N}\right)$$

Where: FC = correction factor; n = number of sampling units; N = potential number of sampling units in the population.

The formulas for finite populations can be considered as generalized and can be used for both cases, since the correction factor becomes insignificant in infinite populations. The purpose of the correction factor is to reduce bias and it exists even in infinite populations

4.1.2 Average

Sample averages are calculated using the expression:

$$\bar{y} = \frac{\sum_{i=1}^{n} \bar{y}_i}{n}$$

Where: $\bar{y}$ = sample mean; $\bar{y}_i$ = mean of sample unit i; n = number of sample units; i = order number of sample unit.

4.1.3 Variance

The sample variance is calculated by:

$$s_x^2 = \frac{\sum_{i=1}^{n} (\bar{y}_i - \bar{y})^2}{n-1}$$

Where: s_y^2 = sample variance; $\bar{y}$ = sample mean; $\bar{y}_i$ = mean of sample unit i; n = number of sample units; i = order number of sample unit.

4.1.4 Sample sufficiency (n)

Sample sufficiency is defined as the number of sample units sufficient to represent the population with the maximum error limit allowed, previously determined in relation to the population mean. The calculation is carried out for an infinite population without the correction factor and for a finite population with the correction factor, as follows:

➤ For infinite population

$$n = \frac{t^2_{(\text{p,GL})} . s^2_{\bar{y}}}{E^2}$$

➤ For finite population

$$n = \frac{N . t^2_{(\text{p,GL})} . s^2_{\bar{y}}}{N . E^2 + t^2_{(\text{p,GL})} . s^2_{\bar{y}}}$$

Where: n = number of sample units sufficient to represent the population; $s^2_{\bar{y}}$ = sample variance; $\bar{y}$ = sample mean; LE = admitted sampling error limit (LE% x Mean); N = total potential number of sampling units; t(p,GL) = Student's t-value for n-1 degrees of freedom (GL) and probability of error p; n = number of sampling units.

4.1.5 Coefficient of variation

The coefficient of variation as a percentage of the mean is calculated using the expression:

$$cv = \frac{100 . s_y}{\bar{y}}$$

Where: CV = coefficient of variation ; s_y = standard deviation; $\bar{y}$ = sample mean.

4.1.6 Variance of the mean

The Variance of the Mean is determined by:

➤ For infinite population

$$S^2_{\bar{y}} = \pm \frac{S^2_{\bar{y}}}{n - 1}$$

➢ For finite population

$$s_{\bar{y}}^2 = \left(\frac{s_y^2}{n-1}\right).\left(\frac{N-n}{N}\right)$$

Where: $S_{\bar{y}}^2$= variance of the mean; S_y^2 = variance; n = number of sample units; N = potential number of sample units.

4.1.7 Standard error of the mean

The Standard Error of the Mean is determined by:

➢ For infinite population

$$S_{\bar{y}} = \pm \frac{s_y}{\sqrt{n}}$$

➢ For finite population

$$S_{\bar{y}} = \pm \frac{s_y}{\sqrt{n}}\sqrt{\left(\frac{N-n}{N}\right)}$$

Where: $S_{\bar{y}}$ = standard error of the mean; s_y = standard deviation; n = number of sampling units; N = potential number of sampling units.

4.1.8 Sample error

The Sampling Error is calculated for absolute and relative values, as follows:

➢ Absolute sampling error (Ea)

$$E_a = \pm\ t_{(p,GL)}.S_{\bar{y}}$$

➢ Relative sampling error (Er)

$$E_r = \pm\ \frac{100.\, t_{(p,GL)}.S_{\bar{y}}}{\bar{y}}$$

Where: Ea, Er = absolute and relative sampling error, respectively; $S_{\bar{y}}$ = standard error of the mean; $t_{(p,GL)}$ = Student's t-value for n-1 degrees of freedom (GL) and probability of error (p); GL = n - 1 (degrees of freedom); p = probability of error (p=1-P).

4.1.9 Confidence interval

$$IC\ [(\bar{y} - t_{(p,GL)}.S_y) \leq \mu \leq (\bar{y} + t_{(p,GL)}.S_{\bar{y}})] = P$$

Where: CI = confidence interval; $\bar{y}$ = sample mean; μ = population mean; $S_{\bar{y}}$ = standard error of the mean; $t_{(p,GL)}$ = Student's t-value for n-1 degrees of freedom (GL) and probability of error (p).

4.1.10 Minimum confidence estimate

$$EMC = \bar{y} - t_{(p,GL)}.S_{\bar{y}}$$

Where: EMC = minimum confidence estimate; $\bar{y}$ = sample mean; $S_{\bar{y}}$ = standard error of the mean; $t_{(p,GL)}$ = Student's t-value for n-1 degrees of freedom (GL) and probability of error (p).

4.2 Restricted sampling

When the forest area is very large, or with different types of forests, there can be a great deal of variation, which may imply a very large number of sampling units for the required precision. In order to reduce the variation between sampling units, sampling is used with division into strata or division into stages, which we have referred to here as multilevel sampling. The division can be by administrative unit, by age, by species, by site quality, in blocks, in stages, etc;

Very large areas can be divided into large 1st level sampling units and these subdivided again into smaller 2nd level sampling units and these can again be subdivided into smaller 3rd level units and so on.

In restricted sampling, there are cases in which the 1st level units have different sizes, so it is necessary to calculate the means and variances considering the proportionality factor, or weight (w_h) of each one, which is calculated by:

$$w_h = A_h / A$$

Where: w_h = weight or proportion of the area of level h over the total; A_h = area of the 1st level unit of order h considered; h = order number of the 1st level unit; A = Total area of the population, obtained by adding up the areas of all the 1st level units.

If the areas of the 1st level units are the same size, the weight w_h will be the same for all 1st level units.

4.2.1 Sampling with a restriction

4.2.1.1 Population division levels for sampling with a restriction

Consider a forest area to be inventoried like the one shown in Figura 23. Dividing the area of the Figura 23 into 9 1st level units, whose order numbers are represented by the letter h.

The total number of 1st level units in the population is represented by the letter H, H=9 in this case, and their order number is h. The number of 2nd level sample units within each 1st level unit is represented by n_h and the order number of each is represented by the letter i.

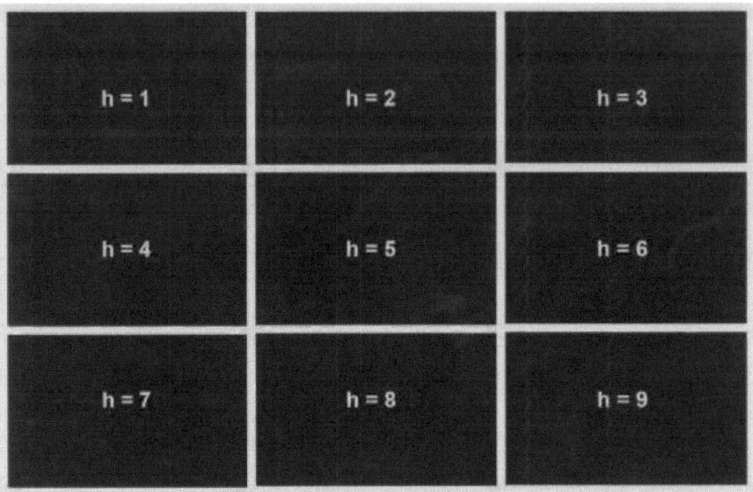

FIGURA 23 - Forest area to be inventoried with representation of 9 1st level units, where h is the order number of the unit in the sampling.

4.2.1.2 Sample mean with a restriction

$$\bar{\bar{y}} = \sum_{h=1}^{H} w_h \, \bar{y}_h$$

Where: $\bar{\bar{y}}$ = average of 1st level units; $\bar{y}_j$ = average of the 1st level unit of order h; w_h = proportion of the area of the 1st level unit of order h over the total area; H = sample number of 1st level units; h = order number of the 1st level unit.

4.2.1.3 1st level unit averages

$$\bar{y}_h = \frac{\sum_{i=1}^{n_h} \bar{y}_{hi}}{n_h}$$

Where: $\bar{y}_{\mathrm{h}}$ = average of the 1st level unit of order h; $\bar{y}_{\mathrm{hi}}$ = average of the 2nd level unit of order i; n_h = number of 2nd level units in the 1st level unit of order h; h = order number of the 1st level unit; i = order number of the 2nd level unit.

4.2.1.4 Sample variance

$$s^2 = \sum_{h=1}^{H} w_h \, s_h^2$$

Where: s^2 = variance; $s_h{}^2$ = variance of the 1st level unit of order h; w_h = proportion of the area of the 1st level unit of order h over the total area; H = sample number of 1st level units; h = order number of the 1st level unit.

4.2.1.5 Sample sufficiency (n)

- Finite population

$$n = \frac{t_{(p,GL)}^2 \sum_{h=1}^{H} w_h . s_h^2}{E^2}$$

- Infinite population

$$n = \frac{N.t_{(p,GL)}^2 \sum_{h=1}^{H} s_h^2}{N.E^2 + t_{(p,GL)}^2 \sum_{h=1}^{H} w_h . s_h^2}$$

Where: $s_h{}^2$ = variance of the 1st level unit of order h; LE = admitted sampling error limit; N = total potential number of sampling units in the population; H = number of 1st level units; w_h = proportion of the area of the 1st level unit of order h over the total area; $t_{(p,\ GL)}$ = Student's t value for n-1 degrees of freedom (GL) and probability of error (p).

4.2.1.6 Allocation of AUs

Once the sample sufficiency (n) has been determined, the AUs can be allocated to each 1st level unit solely by the area proportion (w_h), or by the proportion multiplied by the standard deviation (w_h s_h).

Allocation in proportion to the area of each 1st level UA:

$$n_h = n \ w_h$$

Where: n_h = number of sampling units to be allocated to the 1st level unit of order h; n = number of total sampling units needed in the final inventory; w_h = weight or area proportion of the 1st level unit of order h.

Allocation of sampling units carried out in each 1st level AU by the proportion of area (wh) and standard deviation (sj) of each 1st level unit:

$$n_h = \frac{n \; w_h \, s_h}{s}$$

Where: n_h = number of sampling units to be allocated to the 1st level unit of order h; n = total number of sampling units needed in the definitive inventory; w_h = area proportion of the 1st level AU of order h; s_h = standard deviation of the 1st level AU of order h; s = sampling standard deviation of the stand.

The allocation can also be proportional to the measurement costs, according to the procedures described by Péllico Netto and Brena (1997).

4.2.1.7 Analysis of sampling variance with one restriction

The analysis of variance with a restriction is performed as in Tabela 4.

TABELA 4 - Analysis of sampling variance with one restriction

Variation Factor	Degrees of Freedom	Sum of Squares	Mean Square	F
Between 1st level UAs	H - 1	$\sum_{h=1}^{H} n_h \, (\bar{y}_h - \bar{\bar{y}})^2$	SQe/GLe	QMe/QMd
Within the 1st level UAs	n - H	$\sum_{h=1}^{H} \sum_{i=1}^{n_h} (\bar{y}_{hi} - \bar{y}_h)^2$	SQd/GLd	-
Total	n - 1	$\sum_{h=1}^{H} \sum_{i=1}^{n_h} (\bar{y}_{hi} - \bar{\bar{y}})^2$	'-	-

Interpretation of analysis of variance:

- If the F value is significant: there is a difference between the 1st level units and the division is efficient, reducing the sampling error;
- If the F value is not significant: there is no difference between the 1st level units and division is unnecessary.

4.2.1.8 Coefficient of variation

$$CV = \frac{100\ s}{\bar{\bar{y}}}$$

Where: CV = population coefficient of variation; $\bar{\bar{y}}$ = sample mean; s = sample standard deviation. Statistics with restrictions (2 levels)

4.2.1.9 Variance of the mean

- Infinite population

$$s_{\bar{\bar{y}}}^2 = \sum_{h=1}^{H} \frac{w_h^2 . s_h^2}{n_h}$$

- Finite population

$$s_{\bar{\bar{y}}}^2 = \sum_{h=1}^{H} \frac{w_h^2 . s_h^2}{n_h} - \sum_{h=1}^{m} \frac{w_h . s_h^2}{N}$$

Where: $s_{\bar{\bar{y}}}^2$ = variance of the population mean; $s_h{}^2$ = variance of the 1st level unit of order h; w_h = proportion of the area of the 1st level unit of order h over the total area; H = number of 1st level units; n_h = number of 2nd level units in the 1st level unit of order h; N = potential number of sample units from the whole population; h = order number of the 1st level unit.

4.2.1.10 Sample error

$$E_a = \pm\ t_{(p,GL)}.S_{\bar{\bar{y}}} \qquad E_r = \pm\ \frac{100.t_{(p,GL)}.S_{\bar{\bar{y}}}}{\bar{\bar{y}}}$$

Where: Ea, Er = absolute and relative sampling error, respectively; $\bar{\bar{y}}$ = population mean; $S_{\bar{\bar{y}}}$ = standard error of the population mean; $t_{(p,\ GL)}$ = Student's t-value for n-1 degrees of freedom (GL) and probability of error (p).

4.2.1.11 Confidence interval

$$IC\ [(\bar{\bar{y}} -\ t_{(p,GL)}.s_{\bar{\bar{y}}}) \leq \mu \leq (\bar{\bar{y}} +\ t_{(p,GL)}.s_{\bar{\bar{y}}})] = P$$

Where: CI = confidence interval; $\bar{\bar{y}}$ = sample mean; μ = population mean; $s_{\bar{\bar{y}}}$ = standard error of the population mean; P = probability of confidence for the mean; GL = degrees of freedom = number of strata-1; p = admitted probability of error = 1-P; $t_{(p,\ GL)}$ = Student's t-value for n-1 degrees of freedom (GL) and admitted probability of error (p).

4.2.1.12 Minimum confidence estimate

$$EMC =\ \bar{\bar{y}} -\ t_{(p,GL)}\ s_{\bar{\bar{y}}}$$

Where: CME = minimum confidence estimate; $\bar{\bar{y}}$ = population mean; $s_{\bar{\bar{y}}}$ = standard error of the population mean; p = probability of error; GL = degrees of freedom (number of strata - 1); $t_{(p,GL)}$ = Student's t-value for n-1 degrees of freedom (GL) and probability of error (p).

4.2.2 Sampling with two restrictions

4.2.2.1 Population division levels for sampling with two restrictions

Consider a forest area to be inventoried like the one shown in Figura 24divided into 9 1st level units, whose order numbers are represented by the letter h.

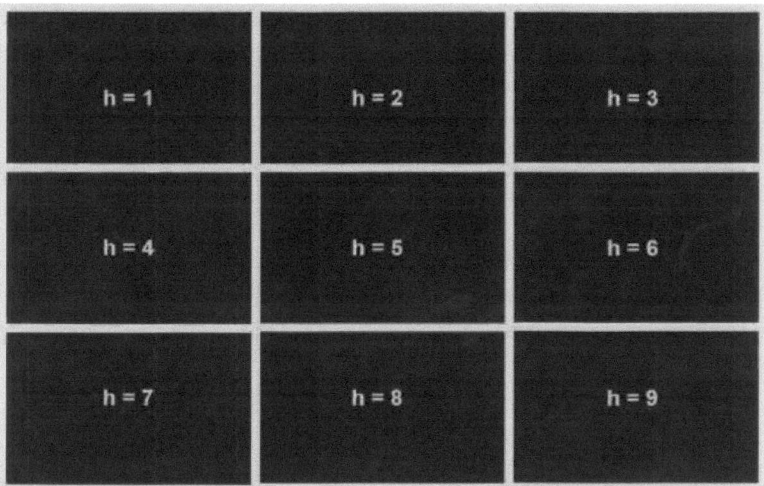

FIGURA 24 - Forest area to be inventoried with representation of 9 1st level units, where h is the order number of the unit in the sampling.

Redividing the 9 1st level units each into another 16 units gives the 2nd level units, represented by the order number i, as in Figura 25.

The number of 1st level units in the entire population is represented by the letter H. The number of 2nd level units within each 1st level unit is represented by n_h . The number of 3rd level UAs within each 2nd level unit is represented by m $._{hi}$

4.2.2.2 Sample average

$$\bar{\bar{y}} = \frac{\sum_{h=1}^{H} w_h \, \bar{y}_h}{\sum_{h=1}^{H} w_h} \qquad ou \qquad \bar{\bar{y}} = \frac{\sum_{h=1}^{H} \sum_{i=1}^{n_h} \sum_{j=1}^{m_{hi}} w_{hij} \; \bar{y}_{hij}}{\sum_{h=1}^{H} \sum_{i=1}^{n_h} \sum_{j=1}^{m_{hi}} w_{hij}}$$

Where: $\bar{\bar{y}}$ = overall sampling average; $\bar{y}_h$ = average of the 1st level unit of order h; w_h = weight or proportion of the area of the 1st level unit of order h over the total area; H = number of 1st level units; h = order number of 1st level units; i = order number of 2nd level units; j = order number of 3rd level units.

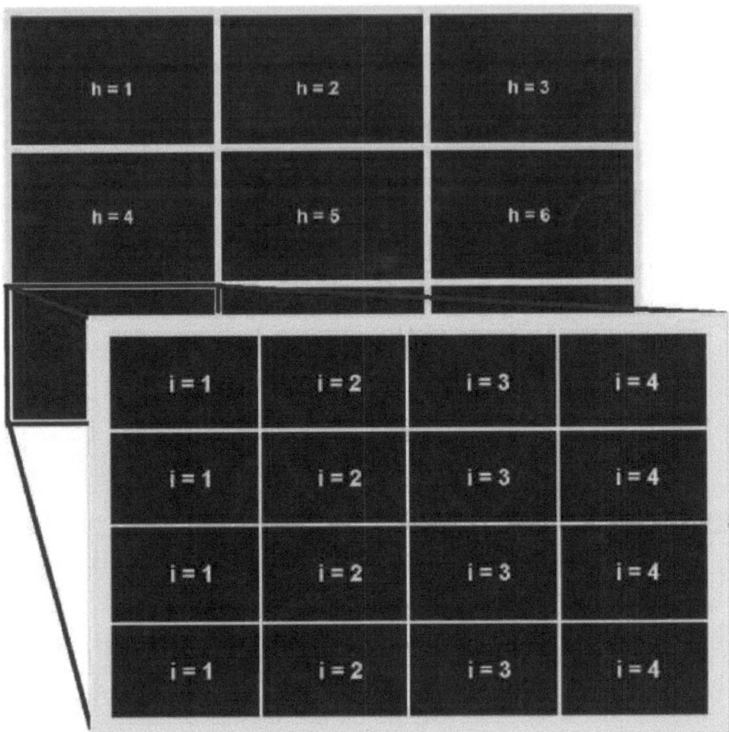

FIGURA 25 - Representation of the 2nd level units (i=1 to 16), of the 1st level unit of order h = 7.

4.2.2.3 Average of 1st level units:

$$\bar{y}_h = \frac{\sum_{i=1}^{n_h} w_{hi}\, \bar{y}_{hi}}{\sum_{i=1}^{n_h} w_{hi}}$$

Where: $\bar{y}_h$ = average of the AU of 1st level of order h; $\bar{y}_i$ = average of the 2nd level AU of order i, within the 1st level AU of order h; w_{hi} = weight or proportion of the area of the 2nd level AU of order i, within the 1st level AU of order h; n_h = number of 2nd level AUs within the 1st level AU of order h; h = order number of the 1st level unit; i = order number of the 2nd level unit within the 1st level AU of order h.

4.2.2.4 Average of 2nd level units:

$$\bar{y}_{hi} = \frac{\sum_{i=1}^{m_{hi}} w_{hij}\, \bar{y}_{hij}}{\sum_{i=1}^{m_{hi}} w_{hij}}$$

Where: $\bar{y}_{hi}$= average of 2nd level AU of order i within unit h; $\bar{y}_{hij}$ = average of 3rd level AU of order j, within 2nd level AU of order i, within 1st level unit of order h; m_{hi} = number of 3rd level units of order j, within 2nd level unit of order i, within 1st level unit of order h; j = order number of 3rd level AUs; i = order number of 2nd level AUs; h = order number of 1st level AUs.

4.2.2.5 Allocation of AUs

4.2.2.5.1 Allocation by area ratio

The sampling units are allocated to each 1st level AU by the area proportion (w_h) of each 1st level AU:

$$n_h = w_h \, . \, n$$

Where: n_h = number of sampling units to be allocated to the 1st level unit of order h; n = number of total sampling units needed in the definitive inventory; w_h = proportion of the area of the 1st level AU of order h.

4.2.2.5.2 Allocation by area ratio and standard deviation

The sampling units are allocated to each 1st level AU by the proportion of area (w_h) and standard deviation (s_h) of each 1st level unit:

$$n_h = \frac{n\ w_h\, s_h}{s}$$

Where: n_h = number of sampling units to be allocated to the 1st level unit of order h; n = number of total sampling units needed in the definitive inventory; w_h = area proportion of the 1st level AU of order h; s_h = standard deviation of the 1st level AU of order h; s = sample standard deviation.

4.2.2.6 Analysis of sampling variance with two restrictions

The analysis of variance with two restrictions is carried out as in Tabela 5.

4.2.2.6.1 Interpretation of the analysis of variance

- If the F value between the 1st level AUs is significant, there is a difference between them and the division by the 1st level AUs was efficient, reducing the variance; otherwise, the division is unnecessary;
- If the F value between the 2nd level AUs is significant, there is a difference between them and the division into 2nd level AUs within the 1st level AUs reduces the variance; otherwise, the sample can be considered as a simple random sample.

otherwise, the sample can be considered simple random, disregarding the division into 2nd level within 1st level.

4.2.2.7 Coefficient of variation

$$CV = \frac{100\,s}{\bar{\bar{y}}}$$

Where: CV = coefficient of variation; $\bar{\bar{y}}$ = overall sample mean; s = sample standard deviation.

TABELA 5 - Analysis of sampling variance with two restrictions

Variation Factor	Degrees of Freedom	Sum of Squares	Mean Square	F
Between 1st level UAs	H -1	$\sum_{h=1}^{H} n_h m_h (\bar{y}_h - \bar{\bar{y}})^2$	$\frac{SQe}{GLe}$	$\frac{QMe}{QMc}$
Between 2nd level UAs within 1st level UAs	H ((n$_h$ /H) -1)	$\sum_{h=1}^{H}\sum_{i=1}^{n_h} n_h(\bar{y}_{hi} - \bar{y}_h)^2$	$\frac{SQc}{GLc}$	$\frac{QMc}{QMp}$
Between 3rd level UAs within 2nd level UAs	H (n$_h$ /H) (m$_{hi}$ /n$_h$ - 1)	$\sum_{h=1}^{H}\sum_{i=1}^{n_h}\sum_{j=1}^{m_{hi}} (\bar{y}_{hij} - \bar{y}_{hi})^2$	$\frac{SQp}{GLp}$	-
Total	H (n$_h$ /H) (m /n$_{hih}$) -1	$\sum_{h=1}^{H}\sum_{i=1}^{n_h}\sum_{j=1}^{m_{hi}} (\bar{y}_{hij} - \bar{\bar{y}})^2$	$\frac{SQt}{GLt}$	-

Where: H = number of 1st level (primary) UAs; n_h = number of 2nd level (secondary) UAs within 1st level UAs; m_{hi} = number of 3rd level (tertiary) UAs within 2nd level UAs; $\bar{\bar{y}}$ = overall sample average; $\bar{y}_h$= average of 1st level AUs of order h; $\bar{y}_{hi}$ = average of the 2nd level AU of order i within the 1st level AU of order h; $\bar{y}_{hij}$ = average of the 3rd level AU of order j within the 2nd level AU of order i, within the 1st level AU of order h; FV = variation factor; GLe, GLc, GLp, GLt = degrees of freedom of the 1st, 2nd, 3rd level AUs and total, respectively; SQe, SQc, SQp, SQt = Sum of squares of the 1st, 2nd and 3rd level AUs and total, respectively; QMe, QMc, QMp, QMt = mean square of the 1st, 2nd and 3rd level AUs and total, respectively; F = Snedecor's F value.

4.2.2.8 Variance of the sample mean

➢ Using the Taylor series method (SAS, 2016):

$$s_{\bar{\bar{y}}}^2 = \sum_{h=1}^{H} s_{\bar{y}_h}^2$$

Where, if $n_h > 1$:

$$s_{\bar{y}}^2 = \frac{n_h(1 - f_h)\ \sum_{i=1}^{n_h}(e_{hi} - \bar{e}_h)^2}{n_h - 1}$$

$$e_{hi} = \frac{\left(\sum_{j=1}^{m_{hi}} w_{hij}\ (y_{hij} - \bar{y})\right)}{w}$$

$$\bar{e}_h = \frac{\sum_{i=1}^{n_h} e_{hi}}{n_h}$$

Where: $s_{\bar{\bar{y}}}^2$ = variance of the sample mean; $s_{\bar{y}_h}^2$ = variance of the mean of the 1st level unit of order h; w_{hij} = weight associated with each observation; H = number of 1st level units; n_h = number of 2nd level units within the 1st level unit of order h; m_{hi} = number of 3rd level units within the 2nd level units; h = order number of 1st level units; i = order number of 2nd level units; j = order number of 3rd level units.

- The conventional method:
  - Infinite population

$$s_{\bar{\bar{y}}}^2 = \sum_{h=1}^{H} \frac{w_h^2 . s_h^2}{n_h}$$

  - Finite population

$$s_{\bar{\bar{y}}}^2 = \sum_{h=1}^{H} \frac{w_h^2 . s_h^2}{n_h} - \sum_{h=1}^{H} \frac{w_h . s_h^2}{N}$$

Where: $s_{\bar{\bar{y}}}^2$ = variance of the sample mean; $s_h{}^2$ = variance of the 1st level AU of order h; w_h = weight of the 1st level AU of order h; H = sample number of 1st level AUs; N = potential number of sample units; h = order number of the 1st level unit.

4.2.2.9 Sample error

$$E_a = \pm\ t_{(p,GL)} . s_{\bar{\bar{y}}} \qquad E_r = \pm\ \frac{100.\, t_{(p,GL)} . s_{\bar{\bar{y}}}}{\bar{\bar{y}}}$$

Where: Ea, Er = absolute and relative sampling error, respectively; $\bar{\bar{y}}$ = sample mean of the population; $s_{\bar{\bar{y}}}$ = standard error of the sample mean; GL = degrees of freedom = number of 1st level AUs minus 1 (H - 1); p = probability of error; $t_{(p,GL)}$ = Student's t-value for H-1 degrees of freedom (GL) and probability of error (p).

4.2.2.10 Confidence interval

The confidence interval is calculated by:

$$IC\ [(\bar{\bar{y}} - t_{(p,GL)} . s_{\bar{\bar{y}}}) \leq \mu \leq (\bar{\bar{y}} + t_{(p,GL)} . s_{\bar{\bar{y}}})] = P$$

Where: CI = confidence interval; μ = true population mean; $s_{\bar{\bar{y}}}$ = standard error of the mean; GL = degrees of freedom = number of strata - 1; p = probability of error = 1 - P; $t_{(p,GL)}$ = Student's t-value for H-1 degrees of freedom (GL) and probability of error (p).

4.2.2.11 Minimum confidence estimate

It is represented by the lower limit of the average:

$$EMC = \bar{\bar{y}} - t_{(p,GL)}\ s_{\bar{\bar{y}}}$$

Where: EMC = minimum confidence estimate; $\bar{\bar{y}}$ = sample mean of the population; $s_{\bar{\bar{y}}}$ = standard error of the population sample mean; p = probability of error; GL = degrees of freedom (number of 1st level units - 1); t(p,GL) = Student's t-value for H-1 degrees of freedom (GL) and probability of error (p).

5. SAMPLING PROCEDURES

They are characterized by the sampling frame, i.e. the combination of the sampling method and the sampling system in the sampling frame.

The main sampling processes are listed below and described in topics 5.1 to 5.5:

- AAS - Simple random sampling;
- SEA - Stratified random sampling;
- AS - Systematic sampling:
    - In Lines;
    - In Tracks;
    - In Networks;
    - With multiple random starts;
- AC - Conglomerate sampling;
- AMS - Multi-stage sampling.

Usually, an inventory is started

5.1 Simple Random Sampling (SRS)

This is the basic method of probabilistic selection. All AUs have the same chance of being selected. The total forest area is divided into N equal Potential Units = the desired shape and size of the Sample Unit and n units are randomly selected to make up the sample.

The statistics used with simple random sampling are those of the unrestricted sampling systems described in Chapter 4.

The distribution of sampling units can be done using the ordination method or the base line method.

There are mainly two methods for randomizing the sampling units in the area to be inventoried: ordination and baseline.

5.1.1 Sorting method (Figura 26)

It is carried out in the following stages:

1. The total area is divided into squares or rectangles equal to the area of the AU, which will form the potential units (N);
2. The potential units are ordered from 1 to N;
3. Leave out the units on the edge;

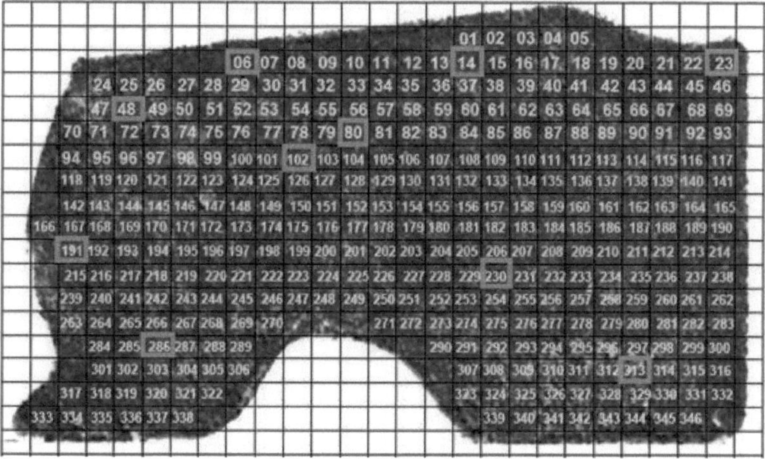

FIGURA 26 - Dividing the forest area into N potential plots to randomize the sampling units.

4. N Sample Units (AU) are drawn from the N potential units.
5. In the example on the next slide, there are 346 potential units that are not on the border;

6. Randomization can be carried out in Excel using the =Randombetween(1;346) function;
7. The following 7 sampling units were drawn for the pilot inventory: 286, 230, 14, 80, 102, 6, 48, 191, 313 e 23.

5.1.2 Base line method (Figura 27)

In this method, the sequence of activities is as follows:

1. A base line is drawn across the entire area to be inventoried;
2. Measure the length of the base line over the area to be inventoried;
3. The border on both sides is deducted from the length of the base line; the size of the AU in the direction of the base line is also reduced;
4. The starting point on the base line will be equal to the border plus the size of the AU (10m + 25m = 22.5m);
5. In Example da, the base line is 705m
6. (705 - 2x10m border - 25m AU = 660),
7. The distance to be traveled from the starting point to the access point for the UA is drawn [=Aleatórioentre(22,5;660) - it gave 450m in this example];
8. A draw is made as to which side of the line the UA will be located (right);
9. The distance from the access point on the base line to the edge of the forest (160 m) is measured;

10. The ½ of the width of the AU (10m) and the 10m border are deducted, leaving 140m;
11. The distance from the access point on the base line is drawn from ½ the width of the AU (10m) to the center of the AU, as follows: [=Sortbetween(10;140) - 85m].

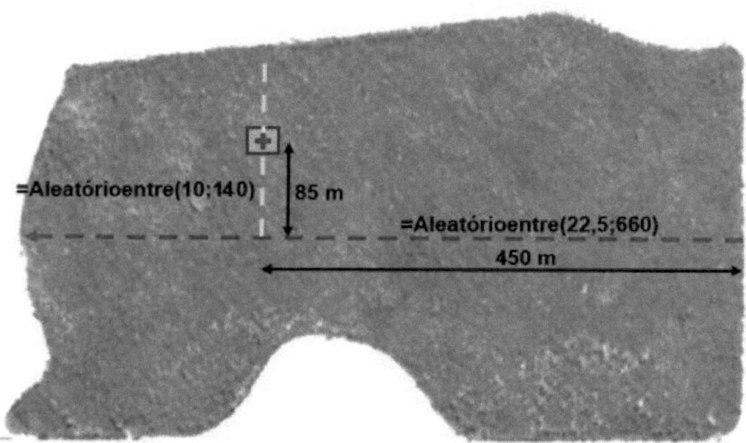

FIGURA 27 - Randomization of sampling units using the base line method.

5.1.3 Example of Simple Random Sampling

Assuming that the area of the forest stand shown in Figura 26 is 18 hectares and it was decided to sample 10 plots in the pilot inventory, whose volumes per hectare per plot are shown in Tabela 6.

The sampling error limit (LE%) accepted in the inventory is 10% of the mean, for a probability of 95% confidence in the mean; and, in absolute terms, the sampling error (E) is calculated by:

$$E = LE\% \,.\, Average = 10\% \,.\, 205.0\ m^3/ha = 20.5\ m^3/ha$$

TABELA 6 - Volume per plot, average and variance of the pilot inventory

Tranche	Volume (m^3/ha)
6	154,0
14	226,0
23	208,0
48	219,0
80	183,0
102	275,0
191	287,0
230	161,0
286	143,0
313	194,0
Average	205,0
Variance	2361,8

Whereas:

- Forest area = 18.0 ha;
- Area of the plots = 500 m^2;
- Potential N = 360 plots;
- Sampling intensity = 2.8% = finite population;
- Degrees of Freedom (GL): 10 plots - 1 = 9 degrees of freedom;
- Value of $t_{(0,95;9)}$ = 2.2622.

Using the equation for a finite population to determine the number of sample units sufficient for the required probability, we have:

$$n = N \,.\, t^2 \,.\, S^2 / (N \,.\, E^2 + t^2 \,.\, S^2)$$

$$n = 360 \,.\, 2.2622^2 \,.\, 2361.8 / (360 \,.\, 21.66^2 + 2.2622^2 \,.\, 2361.8)$$

$$n = 26.6 \sim 27\ AU$$

This results in 27 sampling units (AU) in this first approximation (Tabela 7), and the value should always be approximated upwards. After the first approximation, find the value of t for GL = 27 -1 = 26 degrees of freedom and repeat the calculation of n with the new value of $t_{(p;\ GL)}$ until the value stabilizes. The result for the approximations until the value of the number of units needed (n) in the final inventory stabilizes is shown in Tabela 7for this case.

TABELA 7 - Determining the number of sample units (n) sufficient to represent the population with 95% confidence .

Getting closer	Degrees of freedom	$t_{(p;GL)}$	n enough
1ª	9	2,2622	26,6
2ª	26	2,0555	22,3
3ª	22	2,0739	22,7
4ª	22	2,0739	22,7
No. of sampling units to be inventoried =		23	

The wood stocks per sampling unit in the definitive inventory are shown in Tabela 8. The statistical results of the sampling are presented below.

5.1.3.1 Sampling intensity (SI)

Sampling Intensity (SI) is represented by the percentage of the forest area under study that was sampled, and is calculated by:

$$AI = 100\ .\ n\ .\ a\ /\ A$$

$$IA = 100\ .\ 23\ .\ 500m^2\ /\ (18ha\ .\ 10000) = 6.39\%$$

Where: AI = Sampling Intensity in percentage; n = number of sampling units; a = area of the sampling unit in square meters; A = area of the population in square meters.

Sampling intensity is greater than 2% and the population is finite. Finite populations need to have their statistics corrected to reduce bias, using a multiplicative correction factor (CF) calculated by the expression:

$$FC = \left(\frac{N-n}{N}\right)$$

FC = (360-23) / 360 = 0.936111

Where: FC = correction factor; n = number of sampling units; N = potential number of sampling units in the population.

TABELA 8 - Average volume per plot in the definitive inventory

n	Tranche	Volume ($\bar{y}_i$) (m³/ha)	$(\bar{y}_i - \bar{y})^2$
1	6	154,0	3768,9
2	14	226,0	112,5
3	23	208,0	54,6
4	48	219,0	13,0
5	80	183,0	1049,2
6	102	275,0	3553,2
7	191	287,0	5127,8
8	230	161,0	2958,4
9	286	143,0	5240,5
10	313	194,0	457,6
11	24	263,0	2266,6
12	37	257,0	1731,3
13	56	225,0	92,3
14	138	228,0	159,0
15	244	275,0	3553,2
16	253	195,0	415,8
17	199	177,0	1473,9
18	164	201,0	207,1
19	155	165,0	2539,3
20	262	250,0	1197,8
21	31	158,0	3293,8
22	274	275,0	3553,2
23	113	235,0	384,5
Sum		4954,0	43203,5
Average ($\bar{y}$)		215,4	-

5.1.3.2 Sample mean ($\bar{y}$)

The sample mean is calculated using the expression:

$$\bar{y} = \frac{\sum_{i=1}^{n} \bar{y}_i}{n}$$

$\bar{y}$ *= 4954.0 / 23 = 215.4 m³/ha*

Where: $\bar{y}$ = sample mean; $\bar{y}_i$ = mean of sample unit i; n = number of sample units; i = order number of sample unit.

5.1.3.3 Variance ($s_y{}^2$)

The sample variance is calculated by:

$$s_{\bar{y}}^2 = \frac{\sum_{i=1}^{n}(\bar{y}_i - \bar{y})^2}{n-1}$$

$s_y{}^2 = 43203.5 / (23\text{-}1) = 1963.8\ (m^3/ha)^2$

Where: $s_{\bar{y}}^2$ = sample variance; $\bar{y}$ = sample mean; $\bar{y}_i$ = mean of sample unit i; n = number of sample units; i = order number of sample unit.

5.1.3.4 Standard Deviation (s)$_y$

It is calculated as the square root of the variance:

$$s_y = \sqrt{\ } s_y{}^2$$

$$s_y = \sqrt{1963.8(m^3/ha)^2} = 44.31\ m^3/ha$$

Where: s_y = Standard Deviation in m^3/ha; $s_y{}^2$ = Sample Variance in $(m^3/ha)^2$.

5.1.3.5 Coefficient of variation (CV)

The coefficient of variation as a percentage of the mean is calculated using the expression:

$$cv = \frac{100. s_y}{\bar{y}}$$

$CV = 100.\ 44.31\ m^3/ha / 215.4\ m^3/ha = 20.6\%$

Where: CV = coefficient of variation; s_y = standard deviation; $\bar{y}$ = sample mean.

5.1.3.6 Variance of the mean ($s_{\bar{y}}^2$)

The Variance of the Mean is determined by:

$$s_{\bar{y}}^2 = \left(\frac{s_y^2}{n-1}\right) \cdot \left(\frac{N-n}{N}\right)$$

$S_{\bar{y}}^2$ = *(1963.8 (m³/ha)²/(23-1)).((360-23)/360) = 83.56 (m³/ha)²*

Where : $s_{\bar{y}}^2$= variance of the mean; s_y^2 = variance; n = number of sampling units; N = potential number of sampling units.

5.1.3.7 Standard error of the mean ($S_{\bar{y}}$)

The Standard Error of the Mean is determined by the square root of the variance of the mean:

$$s_{\bar{y}} = \pm\sqrt{s_{\bar{y}}^2}$$

$$s_{\bar{y}} = \pm\sqrt{83,56\ (m^3/ha)^2} = \pm 9.14\ m^3/ha$$

Where: $s_{\bar{y}}$ = standard error of the mean; $s_{\bar{y}}^2$= variance of the mean.

5.1.3.8 Sample error

The Sampling Error is calculated for absolute and relative values, as follows:

- Absolute sampling error (Ea)

$$E_a = \pm\ t_{(p,GL)} . s_{\bar{y}}$$

Where $t_{(0,05;\ 23-1)}$ = 2.0739

Ea = ± 2.0739 . 9.14 m³/ha = ± 18.5 m³/ha

- Relative sampling error (Er)

$$E_r = \pm\ \frac{100.\, t_{(p,GL)} . s_{\bar{y}}}{\bar{y}}$$

*Er = ± 100 * 18.5 m³/ha / 215.4 m³/ha = ± 8.6%*

Where: Ea, Er = absolute and relative sampling error, respectively; $\bar{y}$ = sample mean; $s_{\bar{y}}$ = standard error of the mean; GL = n - 1 (degrees of freedom); p = probability of error; $t_{(p,GL)}$ = Student's t-value for n-1 degrees of freedom (GL) and probability of error.

5.1.3.9 Confidence interval (CI)

The Confidence Interval is associated with the probability of confidence for the sample mean, which is the measure of confidence that the true population mean is contained within the confidence interval, i.e. that the population parameter is within it. It is calculated as:

$$IC\ [(\bar{y} - t_{(p,GL)}.s_{\bar{y}}) \leq \mu \leq (\bar{y} + t_{(p,GL)}.s_{\bar{y}})] = P$$

CI [196.9 m³/ha ≤ μ ≤ 233.9 m³/ha] = 95%

Where: CI = confidence interval; μ = population mean; $\bar{y}$ = sample mean; $s_{\bar{y}}$ = standard error of the mean; P = probability of confidence for the mean; GL = degrees of freedom = number of sample units - 1; $t_{(p,\ GL)}$ = Student's t-value for n-1 degrees of freedom (GL) and probability of error (p = 1 - P).

5.1.3.10 Minimum confidence estimate (MCE)

$$EMC = \bar{y} - t_{(p,GL)}.S_{\bar{y}}$$

EMC = 215.4 m³/ha - 18.5 m³/ha = 196.9 m³/ha

Where: EMC = minimum confidence estimate; $\bar{y}$ = sample mean; $s_{\bar{y}}$ = standard error of the mean; P = probability of confidence for the mean; GL = degrees of freedom (number of sample units - 1); t(p,GL) = Student's t-value for n-1 degrees of freedom (GL) and probability of error (p = 1 - P).

5.1.4 SAS program for analysis of the final sample

```
*SAS Program - Simple Random Sampling;
*r = n/N = 0.0639;
data data;
input UA Parcel Volume;
datalines;
```

```
1 6 154.0
2 14 226.0
3 23 208.0
4 48 219.0
5 80 183.0
6 102 275.0
7 191 287.0
8 230 161.0
9 286 143.0
10 313 194.0
11 24 263.0
12 37 257.0
13 56 225.0
14 138 228.0
15 244 275.0
16 253 195.0
17 199 177.0
18 164 201.0
19 155 165.0
20 262 250.0
21 31 158.0
22 274 275.0
23 113 235.0
;
Proc surveymeans data=data r=0.0639;
*r=n/N;
     var Volume;
run;
```

5.2 Stratified Random Sampling (SRS)

Stratified random sampling is the sampling method in which the area to be inventoried is divided horizontally according to the age, site, species or physiognomy of each subdivision, known as the horizontal vegetation stratum.

Its use is recommended whenever the forest to be inventoried has groups of stands of different types (strata) defined by different ages, different sites, different species or different physiognomies.

Subdivisions or plots with the same characteristics now represent a class and can be treated together to form a single stratum;

Stratification can simply be administrative. The set of strata makes up the forest being inventoried. The advantage over SSA is that SEA allows independent information to be obtained for each stratum and the stratified sampling error is generally smaller for the same sampling intensity due to the stratification itself, requiring a smaller number of sampling units for the same precision.

Stratified random sampling uses the equations for sampling with restrictions, described in Chapter 4.

The population must first be divided into strata and the sampling units of the pilot inventory must be drawn from each stratum and the preliminary mean and variance obtained in order to calculate the sampling sufficiency of the definitive inventory.

5.2.1 Example of Stratified Random Sampling

5.2.1.1 Statement

Suppose we have two areas with different ages and we stratify the population by these two ages. The area to be inventoried is a *Pinus taeda* forest with a total of 34.92 ha, of which 13.90 ha are 35 years old and 21.02 ha are 25 years old, called strata 1 and 2, respectively.

In the pilot inventory it was decided to measure 10 sampling units (AU), 4 in stratum 1 and 6 in stratum 2, according to the area proportion of each stratum. The potential number of sampling units in the population is 34.9 ha / 0.05 ha = 698 potential units. The Sampling Intensity (SI) was calculated at 1.4% (=100.10/698), considering the population to be infinite at this stage.

Consider the information listed below:

- Planted species: *Pinus* taeda;
- Total forest stand area: 34.92 ha;
- Area of 35-year-old stratum 1: 13.90 ha;
- Area of 25-year-old stratum 2: 21.02 ha;
- Area of the sampling unit (AU): 500 m^2;
- Potential number of population plots: 698.4 potential units;
- Potential number of plots in stratum 1: 278 potential units;
- Potential number of plots in stratum 2: 420.4 potential units;
- Proportion of area in stratum 1: 0.398052692;
- Proportion of area in stratum 2: 0.601947308;
- Volume equation: $v = b0 + b1 * d^2h$;
- Probability of confidence for the mean: 95%.

The results of the pilot inventory are shown in Tabela 9 with means and variances.

Since the probability of confidence for the average is 95%, the relative sampling error accepted is (2 x (100%-95%)) = 10% and the absolute sampling error accepted is 64.989 m^3/ha . Next, the sample sufficiency (n) for the definitive inventory was determined, as shown in Tabela 10using the equation for infinite populations, as follows:

$$n = t^2 . s^2 / LE^2,$$

which in the 1st approximation (Tabela 10) resulted in:

$$n = 2.262^2 \,.\, 16505.40(m^3/ha)^2 / 64.989m^3/ha = 20.6$$

Where: $t_{(p;GL)}$ = Student's t for a probability (p) of 95% and 9 degrees of freedom (GL); s^2 = stratified variance of the population; LE = admitted absolute sampling error limit.

TABELA 9 - Results of the pilot inventory with determination of means and variances

Stratum (h)	UA (i)	d (cm)	h (m)	v (m^3)	V = y_i (m^3/ha)	y_h (m^3/ha)	$s_h{}^2$ (m^3/ha)^2	n_h	A_h (ha)	w_h	y (m^3/ha)	s^2 (m^3/ha)^2
1	1	36,76	31,42	1,79	753,22	735,53	4545,42	4	13,90	0,398	649,89	16505,40
1	2	42,38	34,59	2,56	819,51							
1	3	41,15	31,36	2,20	661,22							
1	4	36,68	29,95	1,69	708,19							
2	1	30,46	27,63	1,10	507,04	593,25	24414,23	6	21,02	0,602		
2	2	30,00	26,17	1,01	586,08							
2	3	30,39	27,26	1,07	365,12							
2	4	37,21	28,68	1,64	655,53							
2	5	36,04	29,26	1,61	610,91							
2	6	29,34	26,31	0,99	834,81							
Total	10								34,92			

Where: UA = number of sampling units; d = average diameter of UA; h = average height of UA; v = average individual volume of UA; V = volume per hectare; y_h = average volume of stratum h; $S_h{}^2$ = variance of stratum h; n_h = number of sampling units of stratum h; Aj = area of stratum h; w_h = proportion of area of stratum h; y = average stratified volume of the population per hectare; s^2 = stratified variance of the population.

TABELA 10 - Determining sampling sufficiency for the definitive inventory

Getting closer	Degrees of freedom	$t_{(p;GL)}$	n enough
1ª	9	2,2622	20,6
2ª	20	2,0860	17,4
3ª	17	2,1098	17,8
4ª	17	2,1098	17,8
No. of sampling units required =			18

Even though the population is infinite, the equation can be used for finite populations as in this case, but not the other way around. The sample sufficiency calculation was repeated with each number of degrees of freedom generated, until n stabilized at 18 sample units needed for the definitive inventory.

The number of units needed per stratum (n_h) was allocated proportionally to the area of each stratum ($n_h = w_h \cdot n$), considering the need for 18 plots for the entire forest, as follows:

- $n_1 = w_1 \cdot n = 0.398 \cdot 18 = 7.2$; came up short, resulting in 8 AUs;
- $n_2 = w_2 \cdot n = 0.602 \cdot 18 = 10.8$; approximated upwards, resulting in 11 AUs;
- Adding n1 and n2 gives a total of **19** sampling units needed for the definitive inventory.

The sampling intensity of the definitive inventory was calculated as:

IA = 100 x 19 / 698 = 2.7%, i.e. the population is considered finite at this stage and the sample statistics have to be calculated using the corresponding equations for a finite population. The Correction Factor (CF) used for finite populations is defined by:

$$FC = (N-n) / N$$

$$FC = (698-19) / 698 = 0.97278$$

Where: CF = Correction Factor; N = potential number of population plots; n = number of sample plots.

The results of the final inventory are shown in Figura 28 followed by the sample statistics.

The effectiveness of the stratification was checked using the analysis of variance shown in Tabela 11.

TABELA 11 - Analysis of variance of stratified sampling

Variation Factor	Degrees of Freedom	Sum of Squares	Mean Square	F	PR>F
Between strata	1	18812,1	18812,1	0,88	0.26497 (n.s.)
Within the strata	17	361807,6	21282,8	-	-
Total	18	380619,7	-	-	-

Where: (n.s.) = There was no significant difference between the strata (PR>F=0.12), so statistically stratified sampling did not reduce the sampling error for the same sampling intensity.

Even though there is no difference between the strata, it was decided to keep the stratification because it is represented by two different ages with different growth rates.

The other sample statistics are presented below:

- Stratified average ($\bar{\bar{y}}$):

$\bar{\bar{y}}$ = *734.70 m³/ha*

- Stratified variance ($s^2_{\bar{\bar{y}}}$):

$(s^2_{\bar{\bar{y}}})$ = *21145.5 (m³/ha)²*

- Sampling intensity (SI):

IA = 100 x 19 UA x 500 m² / (10000 . 34.92 ha) = 2.72%

Therefore, the population is considered finite in the final inventory.

- Coefficient of variation (CV):

CV = 20.3%

- Variance of the stratified mean ($S^2_{\bar{\bar{x}}}$):

$$s^2_{\bar{\bar{y}}} = \sum_{h=1}^{H} \frac{w_h^2 . s_h^2}{n_h} - \sum_{h=1}^{m} \frac{w_h . s_h^2}{N}$$

$s^2_{\bar{\bar{y}}}$ =1168.8 - 30.9 = *1137.8 (m³/ha)²*

Estrato (h)	UA (i)	di (cm)	hi (m)	vi (m^3)	Vi (m^3/ha)	y_h	s_h^2	n_h	Área (ha)	w_h	y	s^2	Somas de quadrados			$W_h^2 . S_h^2 / n_h$	$W_h . S_h^2 / N$
													Entre	Dentro	Total		
	1	36,76	31,42	1,7934	753,22									392	343,1		
	2	42,38	34,59	2,561	819,51									2161,9	7194		
	3	41,15	31,36	2,2041	661,22									12499,3	5399,1		
1	4	36,68	29,95	1,6862	708,19	773,02	7074,02	8	13,9	0,3981			11748,3	4203,2	702,8	140,1	4
	5	41,78	29,79	2,1429	685,74									7617	2396,5		
	6	35,84	32,05	1,7176	858,8									7358,6	15401,8		
	7	34,35	30,32	1,4908	805,02									1024,3	4945,8		
	8	35,48	30,53	1,5936	892,44									14261,8	24883,3		
	1	30,46	27,63	1,1023	507,04									40932,2	51828,2		
	2	30	26,17	1,0105	586,08						734,7	21614,01		15196,3	22086,3		
	3	30,39	27,26	1,0739	365,12									118497,2	136585,8		
	4	37,21	28,68	1,6388	655,53									2897	6267,1		
	5	36,04	29,26	1,6077	610,91									9691,1	15322,6		
2	6	29,34	26,31	0,9938	834,81	709,36	31228,95	11	21	0,6019			7063,8	15740,1	10023,7	1028,7	26,9
	7	32,76	31,18	1,4001	812,06									10547,7	5984,7		
	8	34,08	30,33	1,4756	826,35									13687,4	8400,1		
	9	33,5	30,19	1,4078	816,5									11480,8	6692,5		
	10	35,78	30,49	1,6301	945,47									55751,9	44427,1		
	11	33,96	30,06	1,4535	843,03									17867,7	11735,2		
2	19	35,04	29,82	1,57	734,7			19	34,9	1			18812,1	361807,6	380619,7	1168,8	30,9

Onde: UA = número da unidade amostral; d = diâmetro médio da UA; h = altura média da UA; v_i = volume médio individual da UA; V_i = volume por hectare da UA de ordem i; y_h = média volumétrica do estrato h; S_h^2 = variância do estrato h; n_h = número de unidades amostrais do estrato h; A_h = área do estrato h; w_h = proporção de área do estrato h; y = volume médio estratificado por hectare; s^2 = variância

FIGURA 28 - Results of the final inventory of stratified sampling

- Standard error of the stratified mean ($S_{\bar{\bar{x}}}$):

$$s_{\bar{\bar{y}}} = 33.73\ m^3/ha$$

- Value of $t_{(0,05;\ 18)}$ = 2.10092204;
- Sampling error (E):
  - *Absolute sampling error: Ea = 70.86 m³/ha*
    - *Relative sampling error: Er = 9.7%*
- Confidence interval (CI):

  CI = [663.84 m³/ha ≤ μ ≤ 805.56 m³/ha] = 95%
- Minimum confidence estimate (MCE):
  - *For the average: EMC_m = 663.83 m³/ha*
  - *For the total population: EMC_t = 23,181 m³*

5.2.2 SAS program for analysis of the final sample

```
Stratified random sampling;
data data;
input Stratum Parcel Volume Area;
datalines;
1 1 753.2186793 13.9
1 2 819.5138466 13.9
1 3 661.2174177 13.9
1 4 708.1855145 13.9
1 5 685.7422221 13.9
1 6 858.7999303 13.9
1 7 805.0224819 13.9
1 8 892.4405519 13.9
2 1 507.0380862 21.02
2 2 586.0816748 21.02
2 3 365.1209338 21.02
2 4 655.5313567 21.02
2 5 610.9115587 21.02
2 6 834.8146978 21.02
2 7 812.0569627 21.02
2 8 826.3482861 21.02
2 9 816.5037696 21.02
2 10 945.473472 21.02
2 11 843.0253703 21.02
;
proc surveymeans data=data
total=34.92;
     class Stratum;
```

```
var Volume;
Strata Estrato;
Weight Area;
Domain Stratum;
run;
```

5.3 Systematic Sampling (SS)

This is sampling in which the AUs are distributed at regular distances from each other over the entire area to be inventoried. It generally results in greater accuracy of the mean when compared to random sampling, as it covers the entire population uniformly. There is no method for determining the sample variance precisely, only an approximation of the variance. The distribution of the sample units is carried out according to a regular geometric scheme, and can take two forms:

- With only 1 random start;
- With multiple random starts.

5.3.1 Systematic sampling with a single random start

In this type of sampling, only the position of the first unit is randomized. The estimate of variance is not precise, due to the systematization itself. Sampling with a single random start can lead to the recurrence of characteristics of the sampling units, as the location can coincide with the variation in vegetation due to the topography and consequently bias the estimates.

It can be carried out in lines, strips and nets; systematic sampling in nets can be classified as lines with plots aligned in both directions.

It is similar to cluster sampling when you have a single random start, where the lines or bands are similar to clusters made up of smaller units.

We use the equations for sampling with a restriction in which each row or strip represents a 1st level unit and the 2nd level sampling units are the plots in each strip or row.

The difference between lines and strips is that in lines there is spacing between the plots in the line, while in strips there is no distance between the plots within the strip. For both lines and strips, you must first define the distance interval (i) between the lines or strips and, using the forest map, check how many units will be allocated to each line or strip to see if the calculated sampling sufficiency determined with the pilot inventory data will be achieved.

To determine the starting point of the first line or strip, a straight edge line is drawn in the longitudinal direction of the lines or strips. The distance from the internal border line drawn from the forest to the starting point of the first line or strip must be drawn as follows:

- Distance interval (i) between rows or strips: i = 200 m;
- Plots 30 m long (C) and 20 m wide (L);
- Distance interval values from the edge line:
  - Start = (L/2) = 20/2 = 10 m;
  - Final = [i - (L/2)] = 200 - 20/2 = 190 m;
- In Excel or Calc you draw with the function: =Aleatórioentre(10;190)
- In the example in Figura 29 a distance of 50 m to the first sample line was drawn.

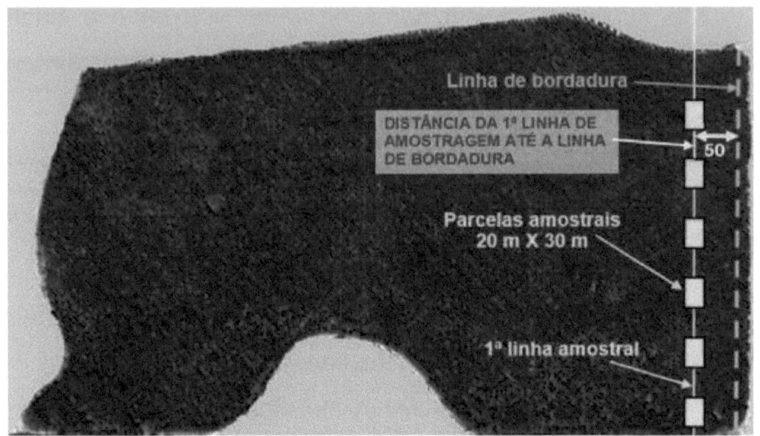

FIGURA 29 - Systematic sampling with a single random start

After the draw (50m), all the other sample lines will be equidistant from each other by 200 meters (Figura 30). The width of the border should be ≥10 m along the entire edge.

In row sampling, the number of rows is determined beforehand by measuring around 3 plots per row of 3 or more rows in the pilot inventory and then adjusting the distance between plots in the rows to obtain the number of plots sufficient for the final inventory.

For the final inventory, measure the total length of the rows, exclude the width of the two borders of each row and divide by the total number of plots needed (n) minus 1, giving the distance between plots:

$$DEL = (CT - 2\,B\,n_L) / (n - 1)$$

Where: DEL = distance between rows; CT = total length of rows; B = width of the forest edge to be avoided; n_L = number of rows in the definitive inventory; n = total number of plots needed in the definitive inventory.

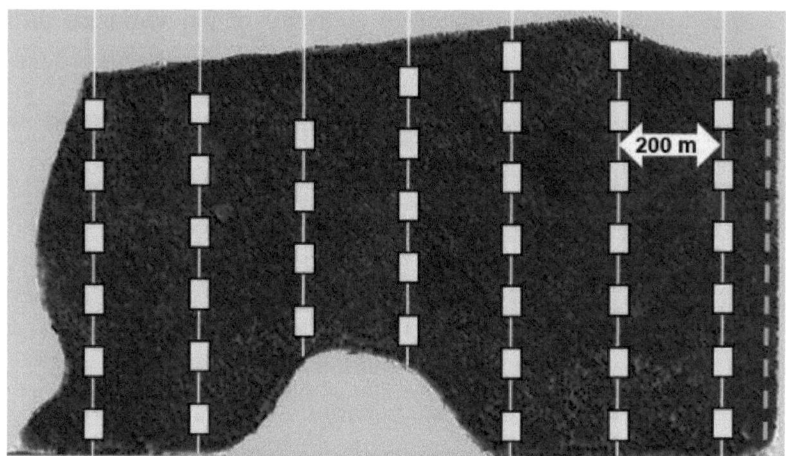

FIGURA 30 - Systematic sampling in lines with an equidistance of 200m between the lines.

In systematic line inventories, the pilot inventory plots have to be placed where they were sampled and are generally not as equidistant as the other plots and will be at an irregular distance from the others in the final inventory.

In strip inventories, the pilot inventory plots have to be entire strips and will not fit into the sampling design of the definitive inventory and will not be equidistant from the others. The right thing to do would be to eliminate the pilot inventory plots, but this would be a waste of time and resources, so it is acceptable for the pilot inventory plots to be displaced from their intended positions in the definitive inventory.

5.3.2 Systematic sampling with multiple random starts

In this type of sampling, the position of each line or band is drawn independently, making the structure represent cluster sampling, providing an accurate estimate of the variance and using the formulas of the restricted sampling system.

It is recommended to use a minimum of 5 sample lines or bands to obtain a more accurate variance.

Systematic sampling with multiple random starts is carried out as follows:

- The distance between lines or stripes is defined;
- The sampling area is divided into bands with a width equal to the distance between the lines or strips;
- The position of each line or sample band is drawn separately from the division limit of each band;
- The draw is made according to the starting and finishing distances, as follows:
  - Start = half the width (L) of the plots;
  - End = distance between rows or strips (D) minus half the width of the plots;
  - In Excel or Calc: =SortBetween(Start;End)
  - Example:
    - L = 20m,
    - D = 200m,
    - Start = 20/2 = 10m,
    - End = 200 - 20/2 = 190 m:
    - =Sortedbetween(10;190)
    - The result is the distance from the line to the band boundary; the draw is made in each band.

The lines are at different distances from the edges of the bands.

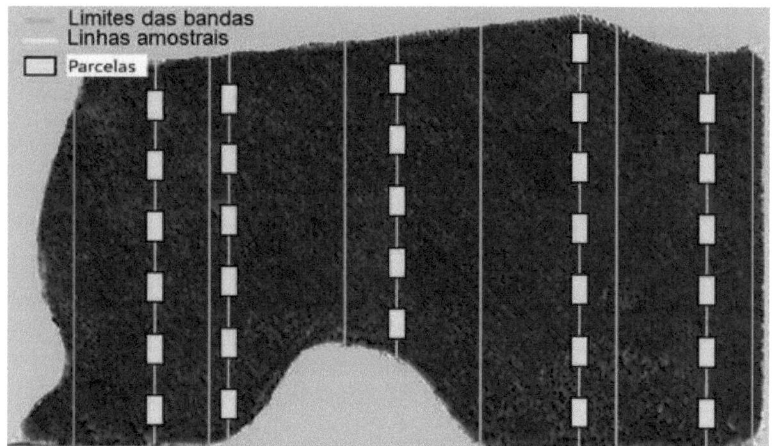

FIGURA 31 - Systematic sampling in lines with multiple random starts.

5.4 Conglomerate Sampling (CA)

Conglomerate sampling is a sampling process in two or more levels where the 1st level sample units are made up of a group with a fixed number of smaller 2nd level units of the same size, which can also be made up of even smaller 3rd level units. Examples of conglomerates are shown in Figura 32.

Various types of design are possible, with the Greek cross format being one of the most widely used.

In this type of sampling, conglomerates generally have a fixed size, but sampling with conglomerates of different sizes, or incomplete conglomerates, can occur. The literature also mentions Bitterlich subunits (DRUSZCZ et al, 2012; RETSLAFF et al, 2014). Sometimes there is more than one type of subunit, especially for assessing regeneration or smaller vegetation (FLORIANO, 2014). The minimum number of subunits to make up a sampling unit is 3, in order to assess the variance within the sampling units.

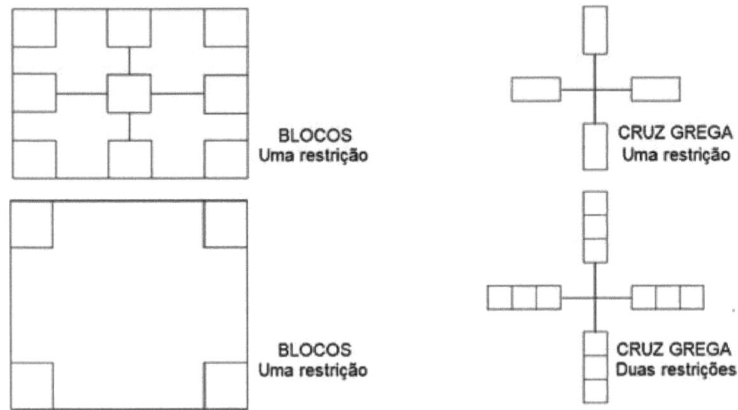

FIGURA 32 - Examples of conglomerates used in forest inventories.

The subunits for assessing regeneration are smaller in size and are generally inserted into the larger subunits. The distance between the subunits of the conglomerates has a high correlation with the variance within and between conglomerates:

- The greater the distance between the plots in the conglomerate, the greater the variance within (Sd^2) and the smaller between the conglomerates (Se^2) up to a point where the variances stabilize;
- The Péllico Netto equation can be used to adjust the distance between plots: $Sd^2 = a + b / D$, where Sd^2 is the variance within the conglomerates, a and b are the parameters of the equation and D is the distance between the plots in the conglomerate;
- Pre-sampling with different distances (D) between plots allows the parameters a and b to be adjusted;

then the equation is plotted and the distance at which Sd^2 stabilizes is found.

The forest to be inventoried can also be stratified by typology, so sampling will have 3 levels (strata, conglomerates within strata and plots within conglomerates), or 4 levels if the units within the conglomerates are divided into smaller plots.

The forest area can be completely divided into small areas corresponding to the conglomerates and the sample conglomerates drawn by lot, or the necessary sample conglomerates can be drawn by coordinates over the area, or by the base line method, or distributed systematically.

5.4.1 Analysis of variance in cluster sampling

The analysis of variance is carried out according to the equations for sampling with restrictions. For cluster sampling with one restriction, it is carried out according to Tabela 12.

If the F value is not significant, there is no significant difference between conglomerates.

Estimates of variance without trend are obtained by (PÉLICO NETTO E BRENA, 1997):

$$s_y^2 = \frac{QM_e + (n_h - 1)\, QM_d}{n_h}$$

$$s_d^2 = QM_d = \frac{\sum_{h=1}^{H} n_h (\bar{y}_h - \bar{\bar{y}})^2}{H-1},$$

$$s_e^2 = \frac{n_h QM_e - QM_d}{n_h},$$

TABELA 12 - Analysis of variance in cluster sampling with one restriction

Variation Factor	Degrees of Freedom	Sum of Squares	Mean Square	F
Between conglomerates	H - 1	$\sum_{h=1}^{H} n_h\,(\bar{y}_h - \bar{\bar{y}})^2$	SQe / GLe	QMe / QMd
Within the 1st level UAs	n - H	$\sum_{h=1}^{H}\sum_{i=1}^{n_h}(\bar{y}_{hi} - \bar{y}_h)^2$	SQd / GLd	-
Total	n - 1	$\sum_{h=1}^{H}\sum_{i=1}^{n_h}(\bar{y}_{hi} - \bar{\bar{y}})^2$	-	-

Where: F = Snedecor's F value; H = sample number of conglomerates; n = total number of subunits from all conglomerates ; n_h = number of subunits per conglomerate; $\bar{y}_{hi}$ = average of subunit i in conglomerate h; $\bar{y}_h$ = mean of conglomerate h; $\bar{\bar{y}}$ = overall sampling mean; GL_e = degrees of freedom between conglomerates ; GL_d = degrees of freedom within conglomerates; SQ_e = sum of squares between conglomerates; SQ_d = sum of squares within conglomerates; QM_e = mean square between conglomerates; QM_d = mean square within conglomerates.

5.4.2 Intra-cluster correlation coefficient (r)

$$r = s_e^{\,2} / (s_e^{\,2} + s_d^{\,2})$$

Where: r = intra-cluster correlation coefficient; $s_e^{\,2}$ = variance between clusters; $s_d^{\,2}$ = variance within clusters.

The closer the intra-cluster correlation coefficient (r) is to zero, the more homogeneous the population is. The population is considered absolutely homogeneous with r=0 and moderately homogeneous with r=0.4.

For cluster sampling, r has an acceptable limit of up to 0.4. If this value is exceeded, there is a high level of heterogeneity in the population and it is necessary to stratify the population into more homogeneous subpopulations.

5.4.3 Example of cluster sampling

Sampling was carried out in Greek cross conglomerates with four subplots, as shown in Figura 33.

Consider the following characteristics of the forest to be inventoried:

- Vegetation: natural broadleaved forest;
- Forest area 1377 ha;
- Sampling method, system and process:
  - Conglomerate random sampling in a Greek cross;
  - Subparcels:
    - 50 m from the center of the conglomerate;
    - Length 100 m;
    - Width 20 m;
    - Plot area 2000 m²;
    - Number of sub-plots per conglomerate 4 sub-plots;
- Sample area of the conglomerates 8000 m²;
- Total area of the conglomerate 90000 m²;
- Potential number of conglomerates 153 conglomerates;
- Probability of confidence for the mean 95%.

5.4.3.1 Pilot Inventory

For the pilot inventory, three conglomerates were sampled.

The data, averages and variances of the pilot inventory are shown in Tabela 13.

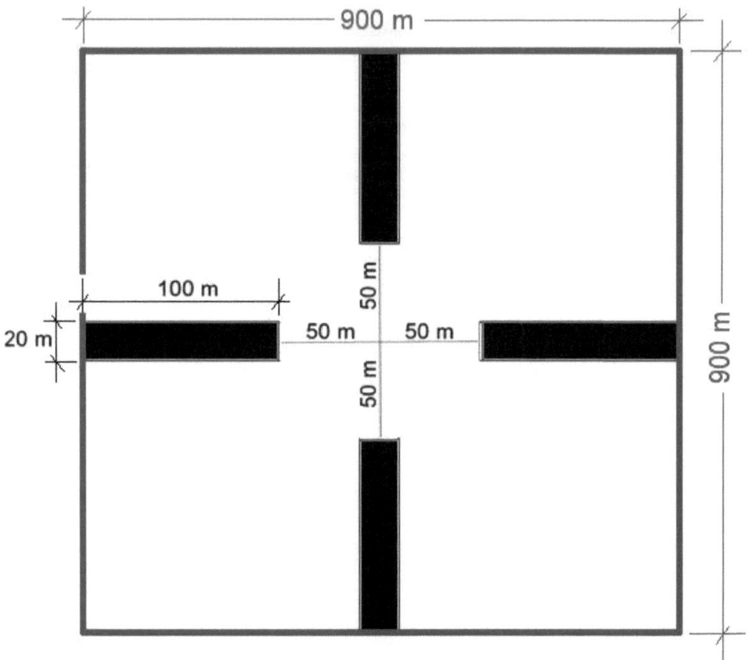

FIGURA 33 - Conglomerate of 9 ha in Greek cross, with one restriction.

TABELA 13 - Results of pilot cluster sampling

Conglomerate (h)	Subunit (i)	y_i (m³/ha)	y_h (m³/ha)	s_h^2 ((m³/ha)²)	n_h	Area (m²)	w_h	y	s²
1	1	139,00	126,50	337,00	4	8000	0,33	125,25	415,86
	2	145,00							
	3	115,00							
	4	107,00							
2	1	120,00	126,75	572,92	4	8000	0,33		
	2	149,00							
	3	96,00							
	4	142,00							
3	1	123,00	122,50	337,67	4	8000	0,33		
	2	113,00							
	3	148,00							
	4	106,00							
Total	12				12	24000			

Where: y = sample mean; y_h = mean of conglomerate of order h; w_h = proportion of conglomerate of order h over the total area; n_h = number of plots in conglomerate h; h = order number of the conglomerate; s_h = variance of conglomerate of order h; s² = sample variance; i = order number of the plot in conglomerate h.

The statistical results were as follows:

- Average (y) = 125.11 m³/ha;
- Variance (s²) = 415.86 (m³/ha)²;
- Probability of confidence for the mean (P) = 95%;
- Relative error limit (LE% = 2 (1-P)) = 10%;
- Absolute error limit (LE = LE% y) = 12.53 m³/ha;
- Potential number of conglomerates (N) = 153;
- sample number of conglomerates (n) = 3;
- Sample Intensity (AI) = 1.96%; implies infinite population.

The calculation of the number of plots needed stabilized at 13 conglomerates in the 5th approximation, as shown in Tabela 14.

TABELA 14 - Determining the number of conglomerates required

Getting closer	Degrees of Freedom (GL)	t (p;gl)	n required = $t^2.S^2/Ea^2$
1ª	2	4,3027	49,1
2ª	49	2,0096	10,7
3ª	10	2,2281	13,2
4ª	13	2,1604	12,4
5ª	12	2,1788	12,6
n needed has stabilized at (approximating upwards) =			13

5.4.3.2 Final Inventory

The data, averages and variances of the final inventory are shown in Tabela 16. The statistical results are presented below:

- Sample number of conglomerates (H) = 13;
- Number of subunits/conglomerate (n_h) = 4;
- Potential number of conglomerates (N_H) = 153;
- Sampling intensity = 8.5% - implies a finite population;
- Average ($\bar{y}$) = 125.11 m³/ha;
- Variance (s^2) = 423.63 (m³/ha)²;
- Standard Deviation (s) = 18.53 m³/ha;
- Coefficient of Variation (CV) = 14.81 %;
- Variance of the mean ($s_{\bar{y}}^2$) = 7.90 (m³/ha)²;
- Standard Error of the Mean ($s_{\bar{y}}$) = 2.81 m³/ha;
- Probability of confidence for the mean (P) = 95%;
- Probability of error (p) = 5%;
- Degrees of freedom (GL) = 12;
- $t_{(p;gl)}$ = 2.1788;
- Absolute Sampling Error (Ea) = +/- 6.12 m³/ha;
- Relative Sampling Error (Er) = +/- 4.90 % - lower than the 10% limit allowed;
- Lower Average Limit (LI) = 119.0 m³/ha;
- Upper Limit of Average (LS) = 159.1 m³/ha;
- Confidence Interval: CI (119.0 m³/ha <= µ <= 159.1 m³/ha) = 95%;
- Minimum Confidence Estimate (MCE) = 119.0 m³/ha

The analysis of variance for cluster sampling (Tabela 15) resulted in a non-significant F value, so there is no difference between the conglomerates.

TABELA 15 - Analysis of variance in cluster sampling

FV	GL	SQ	QM	F	PR>F
Between conglomerates	12	582,30	48,52520	0,1099	0,0054
Within the conglomerates	39	17218,09	441,48944		
Total	51	17103,89	335,370		

- Variances:

$$QMe = 48.525\ (m^3/ha)^2$$

$$QMd = 441.489\ (m^3/ha)^2$$

$$(nh-1)*QMd = 1324.47\ (m^3/ha)^2$$

$$s_y{}^2 = [\ (QMe + (nh-1)*QMd\ ] / n_h = 343.25\ (m^3/ha)^2$$

$$s_e{}^2 = QMe = 48.525\ (m^3/ha)^2$$

$$s_d{}^2 = s_y{}^2 - s_e{}^2 = 294.723\ (m^3/ha)^2$$

- Intra-cluster correlation coefficient (r):

$$r = s_e{}^2 / (s_e{}^2 + s_d{}^2) = 0.141$$

The Intra-Conglomerate Correlation Coefficient showed a value within the acceptable limit (<0.4), there is homogeneity between the conglomerates, it is not necessary to stratify the population, reinforcing the result for the calculated value of F which was not significant, with the interpretation that there is no significant difference between the conglomerates, making it unnecessary to stratify the population.

TABELA 16 - Results of the means and variances of the definitive inventory in conglomerates

Congl. (h)	UA (i)	Y_{hi} (m^3/ha)	Y_h (m^3/ha)	$S_h{}^2$ $(m^3/ha)^2$	n_h	Area (m^2)	w_h	$\overline{y}$	S^2	$(y - y_{hih})^2$	$(y_{-h}\overline{y})^2$	$(y_{-hi}\overline{y})^2$
1	1	139,00	126,50	337,00	4	8000	0,08	125,11	423,63	156,25	7,78	193,07

Congl. (h)	UA (i)	Y_{hi} (m³/ha)	Y_h (m³/ha)	S_h^2 (m³/ha)²	n_h	Area (m²)	w_h	$\bar{y}$	S^2	$(y - y_{hih})^2$	$(y_h - \bar{y})^2$	$(y_{hi} - \bar{y})^2$
	2	145,00								342,25		395,80
	3	115,00								132,25		102,11
	4	107,00								380,25		327,80
2	1	120,00	126,75	572,92	4	8000	0,08			45,56	10,82	26,06
	2	149,00								495,06		570,96
	3	96,00								945,56		847,11
	4	142,00								232,56		285,43
3	1	123,00	122,50	337,67	4	8000	0,08			0,25	27,15	4,43
	2	113,00								90,25		146,54
	3	148,00								650,25		524,17
	4	106,00								272,25		365,01
4	1	137,06	119,69	243,33	4	8000	0,08			301,63	117,19	142,92
	2	128,38								75,47		10,72
	3	109,26								108,84		251,07
	4	104,07								244,06		442,48
5	1	124,49	121,41	20,64	4	8000	0,08			9,50	54,69	0,38
	2	126,11								22,11		1,01
	3	117,59								14,57		56,48
	4	117,44								15,74		58,76
6	1	105,96	126,82	413,36	4	8000	0,08			434,93	11,69	366,54
	2	146,99								407,03		478,94
	3	113,05								189,48		145,33
	4	141,26								208,66		260,98
7	1	103,01	122,06	375,01	4	8000	0,08			362,71	37,21	488,20
	2	110,70								128,94		207,51
	3	128,00								35,34		8,38
	4	146,51								598,05		458,17
8	1	103,59	120,31	127,98	4	8000	0,08			279,39	92,17	462,90
	2	123,29								8,91		3,29
	3	128,00								59,21		8,38
	4	126,34								36,42		1,52
9	1	128,00	125,09	294,36	4	8000	0,08			59,21	0,00	8,38
	2	144,00								561,45		357,01
	3	126,00								32,43		0,80
	4	102,37								321,66		516,89
10	1	133,00	129,00	746,00	4	8000	0,08			161,16	60,68	62,33
	2	156,00								1274,13		954,49
	3	91,00								858,78		1163,16
	4	136,00								246,33		118,70
11	1	93,00	129,00	770,67	4	8000	0,08			745,56	60,68	1030,74
	2	125,00								22,04		0,01
	3	159,00								1497,30		1148,86
	4	139,00								349,50		193,07
12	1	92,00	128,00	800,00	4	8000	0,08			801,17	33,52	1095,95
	2	136,00								246,33		118,70
	3	160,00								1575,69		1217,65
	4	124,00								13,65		1,22
13	1	137,00	129,25	468,25	4	8000	0,08			60,06	68,72	141,49
	2	140,00								115,56		221,86

Congl. (h)	UA (i)	Y_{hi} (m^3/ha)	Y_h (m^3/ha)	$S_h{}^2$ (m^3/ha)^2	n_h	Area (m^2)	w_h	$\bar{y}$	S^2	$(y - y_{hih})^2$	$(y - _h\bar{y})^2$	$(y - _{hi}\bar{y})^2$
	3	97,00								1040,06		789,90
	4	143,00								189,06		320,22
Total	52				13	104000	1,00			17454,94	582,30	17104

Where: $\bar{y}$ = sample mean; y_h = mean of conglomerate of order h; y_{hi} = mean of sub-plot i of conglomerate h; w_h = proportion of conglomerate of order h over the total area; n_h = number of plots in conglomerate h; h = order number of the conglomerate; s_h = variance of conglomerate of order h; s^2 = sampling variance; i = order number of the plot in the conglomerate.

5.4.4 SAS program for analyzing the final inventory

```
* Conglomerate sampling;
data data;
input Conglomerate UA Volume wh;
datalines;
01 1 139.000 0.8
01 2 145.000 0.8
01 3 115.000 0.8
01 4 107.000 0.8
02 1 120.000 0.8
02 2 149.000 0.8
02 3 096.000 0.8
02 4 142.000 0.8
03 1 123.000 0.8
03 2 113.000 0.8
03 3 148.000 0.8
03 4 106.000 0.8
04 1 137.060 0.8
04 2 128.380 0.8
04 3 109.260 0.8
04 4 104.070 0.8
05 1 124.490 0.8
05 2 126.110 0.8
05 3 117.590 0.8
05 4 117.440 0.8
06 1 105.960 0.8
06 2 146.990 0.8
06 3 113.050 0.8
06 4 141.260 0.8
07 1 103.010 0.8
07 2 110.700 0.8
```

```
07 3 128.000 0.8
07 4 146.510 0.8
08 1 103.590 0.8
08 2 123.290 0.8
08 3 128.000 0.8
08 4 126.340 0.8
09 1 128.000 0.8
09 2 144.000 0.8
09 3 126.000 0.8
09 4 102.370 0.8
10 1 133.000 0.8
10 2 156.000 0.8
10 3 091.000 0.8
10 4 136.000 0.8
11 1 093.000 0.8
11 2 125.000 0.8
11 3 159.000 0.8
11 4 139.000 0.8
12 1 092.000 0.8
12 2 136.000 0.8
12 3 160.000 0.8
12 4 124.000 0.8
13 1 137.000 0.8
13 2 140.000 0.8
13 3 097.000 0.8
13 4 143.000 0.8
;
proc surveymeans data=data total=153;
     class Conglomerate;
     var Volume;
     strata Conglomerate;
     Domain Conglomerate;
run;
```

5.5 Multi-stage sampling (MSS)

This is the sampling process in which the entire population area is divided into large sampling units (AUs), which may not be the same size, called 1st stage AUs; these are then redivided into smaller 2nd stage AUs, which can be redivided into even smaller 3rd stage AUs; and so on. A few UAs are then selected at each stage, either randomly or systematically, to make up the pilot inventory sample and determine how many UAs are needed for the desired accuracy.

The distribution of AUs in each stage can be either systematic or random. In the pilot inventory, 3 to 10 AUs can be used in each stage, depending on the size and total (potential) number of AUs in each stage.

Multi-stage sampling is a process used for large areas where access to UAs is difficult. The advantage is the reduction in costs and sampling time for the same accuracy compared to other processes.

5.5.1 Multi-stage sampling statistics

The statistics for this type of sampling can be calculated using the equations for sampling with restrictions, with each stage representing a sampling level.

5.5.1.1 Multi-stage sampling with two restrictions

In this example (Figura 34) the forest area has been divided into u=9 first-stage units (k=first-stage units), which have been subdivided into m=16 subunits each (h=second-stage units) and within each of the 16 subunits there will be n sample plots (i=sample plots).

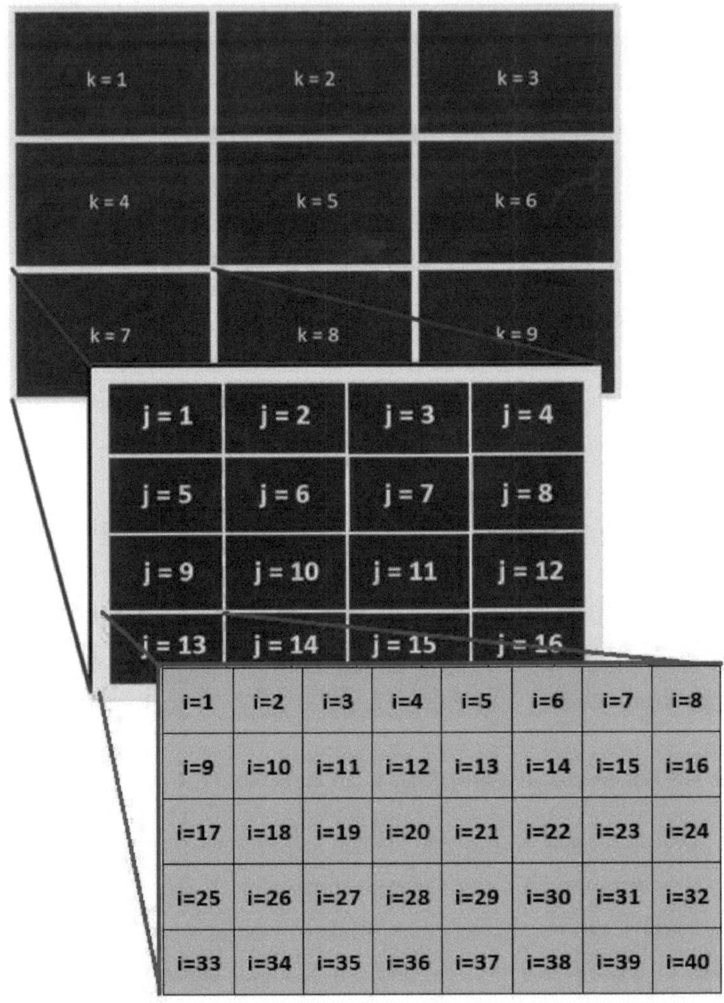

FIGURA 34 - Sampling with two restrictions (1st, 2nd and 3rd level UAs).

The analysis of variance for three-stage sampling is carried out using the equations of the sampling system with two restrictions at 3 levels, as shown in Tabela 17.

TABELA 17 - Analysis of variance with two restrictions (3 stages or levels)

Variation Factor Variation Factor	Degrees of Freedom	Sum of of Squares	Mean Square Mean	F
Between 1st stage UAs	u -1	$\sum_{k=1}^{u} n_j mk\ (\bar{x}_k - \bar{\bar{x}})^2$	$\frac{SQe}{GLe}$	$\frac{QMe}{QMc}$
Between 2nd stage UAs, within 1st stage UAs	u ((m/u) -1)	$\sum_{k=1}^{u}\sum_{j=1}^{m} n_j(\bar{x}_{kj} - \bar{x}_k)^2$	$\frac{SQc}{GLc}$	$\frac{QMc}{QMp}$
Between 3rd stage UAs, within 2nd stage UAs	u (m/u) (n/m-1)	$\sum_{k=1}^{u}\sum_{j=1}^{m}\sum_{i=1}^{n}(\bar{x}_{kji} - \bar{x}_{kj})^2$	$\frac{SQp}{GLp}$	-
Total	u (m/u) (n/m) -1	$\sum_{k=1}^{u}\sum_{j=1}^{m}\sum_{i=1}^{n}(\bar{x}_{kji} - \bar{\bar{x}})^2$	$\frac{SQt}{GLt}$	-

Where: AUs = sample units. $\bar{\bar{x}}$ = overall population mean; $\bar{x}_k$ = average of the 1st stage AU of order k; $\bar{x}_{kj}$ = average of the 2nd stage AU of order h within the 1st stage AU of order k; $\bar{x}_{kji}$ = average of the 3rd stage AU of order i within the 2nd stage AU of order h, within the 1st stage AU of order k; FV = variation factor; GLe, GLc, GLp, GLt = degrees of freedom of the 1st, 2nd, 3rd stage AUs and total, respectively; SQe, SQc, SQp, SQt = Sum of squares of the 1st, 2nd and 3rd stage AUs and total, respectively; QMe, QMc, QMp, QMt = mean square of the 1st, 2nd and 3rd stage AUs and total, respectively; F = Snedecor's F value.

In sampling with two restrictions, consider the following nomenclature for the number of units:

- u = total number of 1st level (primary) UAs;
- m = total number of 2nd level (secondary) UAs;
- n = total number of 3rd level UAs (tertiary = sample plots);
- mk = number of 2nd level UAs within the 1st level UA of order k;
- nh = number of 3rd level UAs of order i, within the 2nd level UA of order h.

6. FOREST MONITORING

Forests are monitored through Continuous Forest Inventories (CFI). Sampling in this case is carried out on multiple correlated occasions.

Sampling is carried out on successive occasions in order to assess variations in tree size and forest structure.

With regard to the occasions on which it is carried out, it can be:

- With total repetition of the AUs on each occasion;
- With partial repetition of the AUs on each occasion;
- With independent UAs on different occasions;
- Double sampling - a variable of interest is estimated based on another variable that is easy to measure;

The main objective of IFCs is to verify changes in the forest over time, mainly in terms of tree growth and wood stock. And it can have the following specific objectives:

- Model tree growth in d, h, h100, g, v, etc...;
- Establishing indices and classifying stand sites;
- Evaluate:

  a) the evolution of the diametric distribution;

  b) the evolution of tree volume and shape;

  c) the evolution of tree density;

  d) the evolution of wood stocks in forest stands;

  e) changes in the profile of the tree trunk.
- Monitor the results of cultural treatments and management actions;

- Providing information for precision forestry;
- To provide information for modeling forest growth and production;
- To provide information for determining the optimum age for technical rotation and cutting cycles.

6.1 Changes in the shape and profile of the trunk

Assessments of changes in the shape and profile of the trunk require cubing by means of appropriate sampling of individual felled trees, which should be carried out at different ages, preferably close to the cuts to give the greatest possible precision in the calculations of volume and assortments to be produced. Trunk analysis can also be used for species that form annual growth rings.

6.2 Frequency of IFCs

The frequency of IFCs depends on:

- Speed of tree growth;
- Cost of sampling on each occasion;
- Type of forest: natural or planted;
- Forest management objective and system:
  - conservation or preservation forest,
  - forest to recover degraded areas,
  - natural production garden forest,
  - high-growth production forest,
  - coppice production forest.

Natural forests with the aim of preservation and conservation can be monitored every 5 years and should include phytosociological surveys. Forests in environmental recovery areas should be monitored over a short period of time in the first few years, after which they can be spaced out to every 5 years. Managed natural forests can be monitored every 2 or 3 years. Planted forests should be surveyed annually.

6.3 Characteristics of IFCs

An error of ±5% around the mean is usually accepted in IFCs. In inventories for growth assessment purposes, the trees in the sampling units should be given permanent numbers at all times. In inventories to provide information for precision forestry, the distribution of sampling units should preferably be systematic. Tree measurements should be taken during the resting period of vegetative growth.

The permanent numbering of the trees makes it possible to evaluate them:

- Ingress (I) - trees that reach the minimum size for measurement.
- Mortality (M) - trees that have died since the last occasion.
- Cutting or thinning (C) - trees that have been harvested since the last occasion.
- Individual growth (ΔY): d, h, g, v.

The IFCs make it possible to evaluate the growth of all the variables measured in the trees and to model their growth curves (Figura 35).

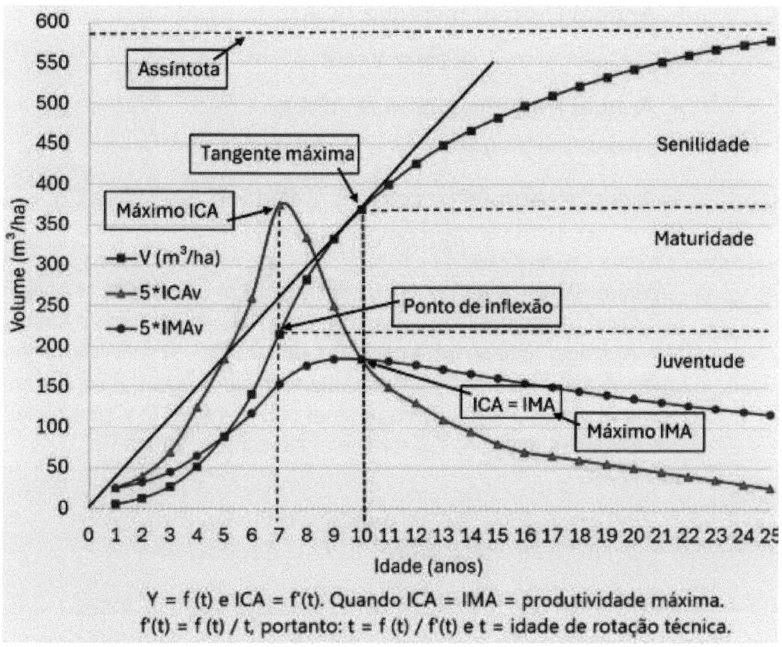

FIGURA 35 - Characteristics of the growth curve forestry.

6.4 Increments

Any dendrometric variable can be analyzed in terms of growth and increments, such as diameter (d), height (h), individual basal area (g), individual volume (v), basal area per hectare (G), volume per hectare (V), crown diameter (CD), etc. Increments are the parts of growth that occur over short periods, usually one year. The most common increments are average annual increment, current annual increment, periodic annual increment and average annual increment at cutting age, calculated as follows:

- Average Annual Increase (IMA):

$$IMA_t = Y_t / t$$

- Annual Current Increase (ICA):

$$ICA_t = Y_t - Y_{t-1}$$

- Annual Periodic Increase (IPA):

$$IPA_t = (Y_t - Y_{t-P}) / P$$

- Average Annual Increment at Cutting Age (IMAIC):

$$IMAIC = Y_r / r$$

Where: IMA_t = average annual increment at age t; ICA_t = Annual Current Increment (ICA); IPA_t = Annual Periodic Increment; IMAIC = Average Annual Increment at Cutting Age; Y_t = value of the variable at age t; Y_{t-1} = value of the variable at age t-1; t = current age in years; r = rotation age (final cut) in years; P = time period considered, in years; Y_{t-P} = value of the variable at age t-P.

6.5 Technical Rotation (RT)

The age at which the Maximum Average Annual Increase in volume per hectare (V) occurs is known as the Technical Rotation age, as it is the age at which the forest's productivity is maximum. At this age, the IMA and ICA equalize and the maximum tangent of the growth curve occurs (Figura 35).

Example: Consider the growth data for a stand shown in Tabela 18. Between the ages of 24 and 25, the Annual Current Increase and the Average Annual Increase in volume per hectare are equal. At this age, the productivity of the stand is maximum and is considered to be the technical rotation age (Figura 36).

TABELA 18 - Growth, increments and technical rotation age of a forest

Age (years)	V (m^3/ha)	ICA (m^3/ha/ year)	IMA (m^3/ha/ year)
	0,2	0,22	0,22
2	2,3	2,12	1,17
3	8,7	6,35	2,90
4	21,1	12,37	5,27
5	40,5	19,40	8,09
6	67,2	26,72	11,20
7	101,0	33,79	14,42
8	141,2	40,19	17,65
9	186,9	45,70	20,76
10	237,1	50,19	23,71
11	290,7	53,63	26,43
12	346,7	56,04	28,89
13	404,2	57,52	31,10
14	462,4	58,14	33,03
15	520,4	58,03	34,69
16	577,7	57,30	36,11
17	633,8	56,05	37,28
18	688,2	54,40	38,23
19	740,6	52,43	38,98
20	790,8	50,22	39,54
21	838,7	47,86	39,94
22	884,1	45,39	40,18
23	926,9	42,87	40,30
24	967,3	40,35	40,30
25	1005,1	37,85	40,21
26	1040,5	35,40	40,02
27	1073,6	33,03	39,76
28	1104,3	30,74	39,44
29	1132,9	28,56	39,06
30	1159,3	26,48	38,64

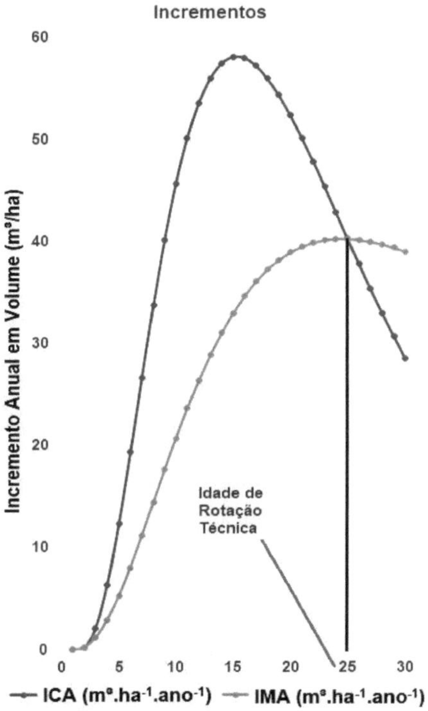

Where: V = total volumetric production of the forest stand per hectare; ICA = Annual Current Increase in volume; IMA = Average Annual Increase in volume.

FIGURA 36 - ICA, IMA and Technical Rotation Age of a forest stand.

6.6 Sampling processes at IFCs

IFCs can be carried out using four different procedures, as follows:

- Sampling with total repetition;
- Sampling with partial repetition;
- Independent sampling;
- Double sampling.

6.6.1 Sampling with total repetition

Total repetition sampling has the following characteristics:

- The sampling units are all permanent;
- All sampling units are measured at each inventory occasion;
- It is ideal for growth analysis;
- The analysis of growth takes into account the covariance between the two occasions;
- It is expensive and not always used by companies due to its high cost.

Example:

In a simple random sampling inventory of a 10-hectare *Pinus elliottii* forest stand, 16 plots measuring 420 m^2 each were used to obtain the results in the table below on two successive occasions. The probability of confidence for the mean was 95%. The two occasions considered here were at 10 and 11 years of age.

The results of the sampling and the increases that occurred are shown in Tabela 19. The statistics for the two occasions and the increases from one to the other are shown in Tabela 20.

TABELA 19 - Results of sampling on two occasions and increases that occurred

Tranche	Volume (m^3/ha)		
no.	10 years	11 years	Increments
1	201,94	247,73	45,80
2	154,45	199,73	45,28
3	190,33	225,01	34,69
4	126,46	164,63	38,17
5	198,11	250,14	52,03
6	189,87	220,76	30,89
7	152,20	181,29	29,09
8	101,41	130,01	28,60
9	202,80	243,70	40,90
10	161,46	217,68	56,22
11	123,23	162,74	39,50
12	111,33	145,82	34,48
13	142,54	190,30	47,76
14	177,65	213,60	35,96
15	131,09	172,07	40,98
16	88,84	119,52	30,68

The equations used for the calculations are those of unrestricted sampling systems for finite populations due to the sampling intensity of more than 2%.

TABELA 20 - Statistics for the two occasions and the increases that occurred

Statistics	10 years	11 years	Increments	Units
Average =	153,36	192,80	39,4	m³/ha
Variance =	1402,43	1706,85	68,26	(m³/ha)²
Standard deviation =	37,45	41,31	8,3	m³/ha
Covariance =	-	-	1520,5	(m³/ha)²
Coefficient of Variation =	24,42	21,43	20,9	%
N potential =	238,10	238,10	238	installments
sample n =	16,00	16,00	16	installments
Sample Intensity =	4,44	4,44	4,44	%
Correction Factor =	0, 9328	0, 9328	0,9328	-
Degrees of freedom (gl) =	15,00	15,00	15	-
Probability of confidence (p) =	0,95	0,95	0,95	-
t (p; gl) =	1,7531	1,7531	1,7531	-
Variance of the mean =	81,76	99,51	3,98	(m³/ha)²
Standard error of the mean =	9,04	9,98	2,0	m³/ha
Absolute sampling error = +/-	15,85	17,49	3,5	m³/ha
Relative sampling error =	10,34	9,07	8,9	%
Minimum confidence estimate =	137,50	175,31	35,9	m³/ha
Confidence interval (CI=95%):	-	-	-	-
- Lower limit of the average =	137,50	175,31	35,94	m³/ha
- Upper limit of the average =	169,21	210,28	42,94	m³/ha

6.6.2 Independent sampling

The characteristics of sampling on multiple independent occasions are as follows:

- The sampling units are all temporary;
- Each occasion is an independent inventory of the same area;
- The trees are not the same on each occasion and therefore it is not a suitable process for growth analysis;

- There is no covariance between the occasions, i.e. Syx=0 and the variance of the mean increment is therefore calculated as the sum of the two occasions;
- The means and variances are obtained in the same way as in total repetition sampling.

The increment between the two independent samples is calculated by:

$$iX = X_2 - X_1$$

Where: iX = increase in variable X; X_1 = value of the variable on the 1st occasion; X_2 = value of the variable on the 2nd occasion.

The variance, standard deviation and coefficient of variation are calculated according to the sampling system used. What changes is the calculation of the variance of the mean, which is no longer reduced by the covariance (Syx), which in this case is equal to zero. The variance of the mean increment is calculated by adding the variances of the means of the two occasions using the equation:

$$S_{\overline{iX}}^2 = S_{\bar{\bar{X}}_1}^2 + S_{\bar{\bar{X}}_2}^2$$

Where: $S_{\bar{\bar{ix}}}^2$ = Variance of the mean increment; $S_{\bar{\bar{X}}_1}^2$ = Variance of the mean on the 1st occasion; $S_{\bar{\bar{X}}_2}^2$ = Variance of the mean on the 2nd occasion.

The standard error of the increment ($S_{\overline{iX}}$) can be calculated simply by taking the square root of the variance of the mean.

The other statistics are calculated according to the sampling system used.

6.6.3 Partial Repeat Sampling

In this type of sampling, some AUs are remediated and other AUs are left unmeasured from one occasion to the next and there are different methods for making estimates and calculating sample and increment statistics.

In the case of inventories of forests planted for timber production, a good practice is to estimate the growth of individual tree dimensions such as diameter, height and volume of the individuals as a function of the dominant height of the AUs expressed in the site index tables and to carry out the statistics as in the total repetition with all the sampling units and their estimated or measured trees.

6.6.3.1 Partial Repeat Sampling with estimates for trees

Traditional books on forest inventories show how to make estimates for the averages of units that have not been sampled, or how to obtain results on the differences from one occasion to another when the units are not all measured on each occasion, disregarding the estimates for individual trees.

However, this is no longer justified with today's computing power, which allows us to accurately estimate the data of each tree from one year to the next based on its growth. By estimating the dimensions of each individual in the unmeasured sampling units, we have the data for all the sampling units tree by tree, which in some units will be estimates and in others the trees will have been measured. The calculations are then carried out as in total repetition sampling. To do this, we first define the proportion of growth in tree diameter and height from one year to the next on all the occasions when the measurement of each sampling unit was repeated, using the equations:

$$pd = (d_{i+1} - d_i) / d_i \text{ e } ph = (h_{i+1} - h_i) / h_i$$

Where: pd = proportion of growth in diameter; ph = proportion of growth in height; d_i = diameter on the 1st occasion considered; d_{i+1} = diameter on the 2nd occasion considered; h_i = height on the 1st occasion considered; h_{i+1} = height on the 2nd occasion considered.

With the age, pd and ph data for each tree from previous inventories, the model is used:

$$pd=A*EXP(-B*t) \text{ and } ph=A*EXP(-B*t)$$

Where: pd, ph = proportion of diameter and height growth, respectively, from age i to age i+1; d_i , h_i = tree diameter and height, respectively, at age i; d_{i+1} , h_{i+1} = tree diameter and height, respectively, at age i+1; t = age at time i; A, B = parameters of the equations to be adjusted;

Then, the diameters and heights at age i+1 are calculated using the equations:

$$d_{i+1} = d_i + d_i *(A*EXP(-B*t)) \text{ and } h_{i+1} = h_i + h_i *(A*EXP(-B*t))$$

Where: d_{i+1} , h_{i+1} = tree diameter and height, respectively, at age i+1; d_i , h_i = tree diameter and height, respectively, at age i; t = age at time i; A, B = parameters of the equations to be adjusted.

6.6.3.2 Estimates of dendrometric variables from one occasion to the next

Estimates of any dendrometric variables (X) from one occasion to the next are made as follows:

- whereas:

$$X_i * pX = X_{i+1} - X_i \ (1)$$

- then:

$$pX = (X_{i+1} - X_i) / X_i \ (2)$$

- or:

$$X_{i+1} - X_i = X_i * pX \ (3)$$

- e:

$$X_{i+1} = X_i + X_i * pX \ (4)$$

- doing:

$$pX = A * EXP(-B*t) \ (5)$$

- and if you substitute 5 for 4, you get:

$$X_{i+1} = X_i + X_i * (A*EXP(-B*t)) \ (6)$$

Where: X_{i+1} = size of the dendrometric variable X at age i+1; X_i = size of the dendrometric variable X at age i; t = age at time i; A, B = parameters of the equations to be adjusted.

The A and B parameters need to be adjusted with data from forest inventories carried out in similar stands, with the same germplasm and the same site and at the different ages for which the estimates are to be made. Alternatively, you could use data from previous years from the same stand, fitting growth curves for the variables of interest and then estimating annual growth. Then, using these estimates, determine the proportion of growth for the desired years and carry out the procedure described in this section.

6.6.4 Example of a continuous forest inventory with partial repetition and estimated trees

This example was carried out with fictitious data and processed in Excel spreadsheets (xlsx) considering unmeasured plots, whose trees had d and h estimated by regression.

The aim was to carry out a continuous forest inventory of a 100-hectare *Pinus elliottii* stand at 13 years of age using stratified random sampling with partial repetition.

To this end, data on diameter (d) and height (h) were collected from 5 strata, with 4 sample plots per stratum, each measuring 295 m². Tree volumes were estimated using a volumetric equation adjusted with data from 24 trees felled in the inventory area.

- The activities to be carried out were as follows:
  Building site index (SI) curves.
  Classifying plots and sites by SI.
  Make estimates for the plots not measured at 13 years of age, since plot 1 of site 1, plot 4 of site 2 and plot 3 of site 4 were not measured at this age.
- Develop a distribution table by diameter class for the age of 13.
- Carry out analysis of variance and calculate the inventory results.

The spreadsheets in the Excel file have the following contents:

1. statement	6. IS	12. V(m³/ha)
2. Trees	7. df	13. DataP
3. h100	8. hf	14. Pilot
4. h100perPlot	9. 13years	15. DataD
5. Eqh100	10. Cubage	16. Definitive
	11. Volume	

6.6.4.1 Spreadsheet: Statement

In Figura 37 shows the "**Statement**" spreadsheet with the general inventory data.

	A	B	C	D	E	F	G
1	**Inventário florestal contínuo com repetição parcial**						
2	Processo de amostragem:	Amostragem aleatória estratificada					
3		em 5 locais.					
4	Idades de medição:	5, 6, 7, 8, 9, 10, 11 e 12 anos					
5	Espaçamento:	3	m² X	2,8	m² =	8,4	m²
6	Parcelas:	5	árv X	7	árv =	294	m²
7	Árvores dominantes de Assmann por parcela:	3	árvores				
8	Número de parcelas por local:	4	parcelas				
9	Local (j)	Área	unid.	Pj			
10	1	17	ha	0,17			
11	2	28	ha	0,28			
12	3	15	ha	0,15			
13	4	14	ha	0,14			
14	5	26	ha	0,26			
15	Total	100	ha	1,00			
16	Probabilidade de confiança para a média:	95%					
17	Número portencial de parcelas:	3401					

FIGURA 37 - Worksheet: Statement - general information about the inventory.

6.6.4.2 Worksheet: Trees

This spreadsheet (Figura 38) contains the data from the measurements of the trees on the plots per site.

	A	B	C	D	E	F
1	Idade	Local	Parcela	Árvore	d	h
2	5	1	1	1	7,0	4,3
3	5	1	1	2	8,2	4,3
4	5	1	1	3	9,2	4,5
5	5	1	1	4	8,5	4,4
6	5	1	1	5	9,0	4,1
7	5	1	1	6	8,4	4,6
...			...		...	
6296	13	5	4	30	35,5	18,1
6297	13	5	4	31	36,4	18,1
6298	13	5	4	32	25,9	16,2
6299	13	5	4	33	38,2	18,1
6300	13	5	4	34	25,9	16,2
6301	13	5	4	35	37,2	18,1

FIGURA 38 - Tree data collected from the plots.

The tree data sheet (Figura 38) collected from the plots contains the following information:

- Age - age of the trees in years;
- Location - forest stratum number;
- Plot - number of the sample plot in the stratum;
- Tree - number of the tree in the plot;
- d - diameter of the tree at a height of 1.3 m, in centimeters;
- h - total height of the tree in meters.

6.6.4.3 Spreadsheet: h100

The data from the trees spreadsheet was copied to the h100 spreadsheet and reordered by Age/Site/Plot/Diameter (the latter from largest to smallest); then the average height of the 3 dominant trees in each plot was calculated. An example was used to determine the dominant height of plot 1, site 1, age 5 years, with the h_{100} of the plot calculated using the formula: =AVERAGE(F2:F4) (Figura 39).

G2 fx =MÉDIA(F2:F4)

	A	B	C	D	E	F	G
1	Idade	Local	Parcela	Árvore	d	h	h100
2	5	1	1	17	11,8	5,8	5,43
3	5	1	1	16	10,0	5,9	
4	5	1	1	30	9,8	4,6	
5	5	1	1	33	9,7	4,6	
6	5	1	1	15	9,5	4,6	
				...			
34	5	1	1	23	6,6	4,0	
35	5	1	1	28	6,0	3,7	
36	5	1	1	18	4,6	3,4	

FIGURA 39 - Example of determining the dominant height .

The h100 column should contain the average of the 3 thickest trees in the plot only in the first row of each plot and the other cells of the trees in this plot will be blank in column G. The determination of h_{100} must be carried out on all plots - note that the formula in column G can be copied and pasted into the first row of each plot.

6.6.4.4 Spreadsheet: hdom

The data from the h100 spreadsheet must be copied and only the values pasted into the hdom spreadsheet. Next, the data is sorted by h100 column (column G) from largest to smallest; then the rows with no dominant height are deleted, leaving only the rows with the dominant heights of the plots. Then the **Tree, d, h** columns are deleted. The following sequence of operations must be carried out on this spreadsheet:

1) Copy the data from worksheet h100 and "paste special" only the "values" into this worksheet;
2) Sort the spreadsheet by h100 from highest to lowest value;
3) Delete the lines without h100;
4) Sort by Age/Location/Parcel;
5) Delete columns: Tree, d, h ;

The result is the spreadsheet in "Appendix B - h_{100} of the plots".

6.6.4.5 Spreadsheet: Eqh100

This spreadsheet shows the fit of the Chapman-Richards model to estimate h100 as a function of age.

The data from the hdom worksheet must be copied and pasted into the Eqh100 worksheet (Figura 40).

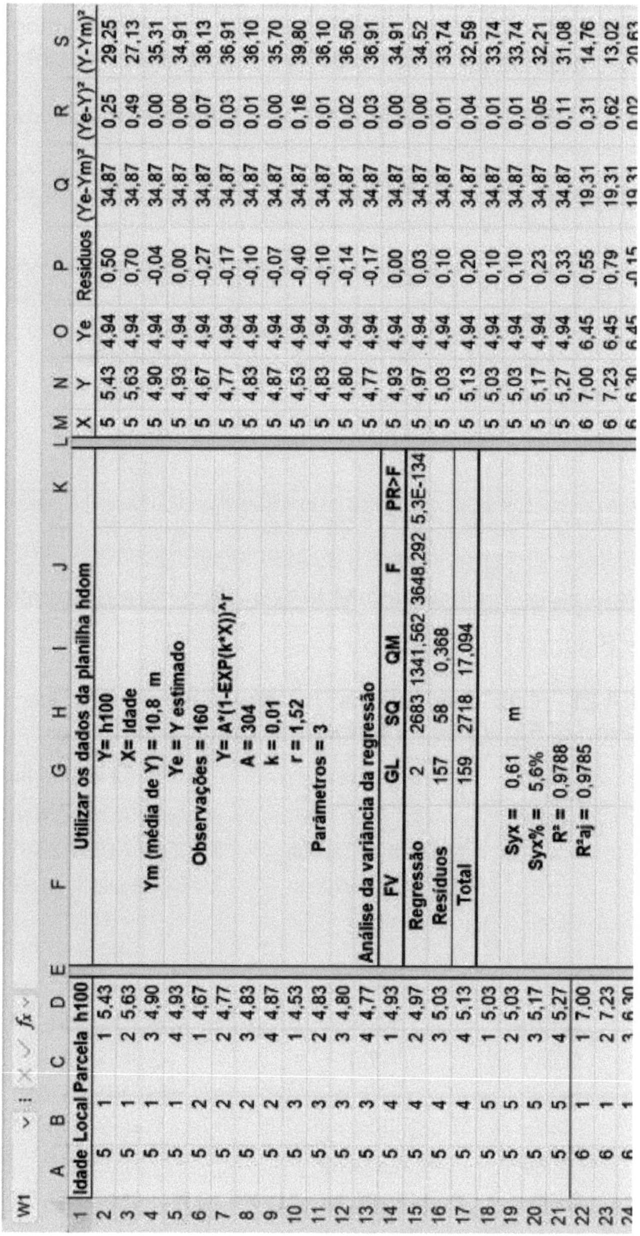

	A	B	C	D
1	Idade	Local	Parcela	h100
2	5	1	1	5,43
3	5	1	2	5,63
4	5	1	3	4,90
5	5	1	4	4,93
6	5	2	1	4,67
7	5	2	2	4,77
8	5	2	3	4,83
9	5	2	4	4,87
10	5	3	1	4,53
11	5	3	2	4,83
12	5	3	3	4,80
13	5	3	4	4,77
14	5	4	1	4,93
15	5	4	2	4,97
16	5	4	3	5,03
17	5	4	4	5,13
18	5	5	1	5,03
19	5	5	2	5,03
20	5	5	3	5,17
21	5	5	4	5,27
22	6	1	1	7,00
23	6	1	2	7,23
24	6	1	3	6,30

Utilizar os dados da planilha hdom

Y= h100

X= Idade

Ym (média de Y) = 10,8 m

Ye = Y estimado

Observações = 160

Y= A*(1-EXP(k*X))^r

A = 304

k = 0,01

r = 1,52

Parâmetros = 3

Análise da variância da regressão

FV	GL	SQ	QM	F	PR>F
Regressão	2	2683	1341,562	3648,292	5,3E-134
Resíduos	157	58	0,368		
Total	159	2718	17,094		

Syx = 0,61 m

Syx% = 5,6%

R² = 0,9788

R²aj = 0,9785

X	Y	Ye	Residuos	(Ye-Ym)²	(Ye-Y)²	(Y-Ym)²
5	5,43	4,94	0,50	34,87	0,25	29,25
5	5,63	4,94	0,70	34,87	0,49	27,13
5	4,90	4,94	-0,04	34,87	0,00	35,31
5	4,93	4,94	0,00	34,87	0,00	34,91
5	4,67	4,94	-0,27	34,87	0,07	38,13
5	4,77	4,94	-0,17	34,87	0,03	36,91
5	4,83	4,94	-0,10	34,87	0,01	36,10
5	4,87	4,94	-0,07	34,87	0,00	35,70
5	4,53	4,94	-0,40	34,87	0,16	39,80
5	4,83	4,94	-0,10	34,87	0,01	36,10
5	4,80	4,94	-0,14	34,87	0,02	36,50
5	4,77	4,94	-0,17	34,87	0,03	36,91
5	4,93	4,94	0,00	34,87	0,00	34,91
5	4,97	4,94	0,03	34,87	0,00	34,52
5	5,03	4,94	0,10	34,87	0,01	33,74
5	5,13	4,94	0,20	34,87	0,04	32,59
5	5,03	4,94	0,10	34,87	0,01	33,74
5	5,03	4,94	0,10	34,87	0,01	33,74
5	5,17	4,94	0,23	34,87	0,05	32,21
5	5,27	4,94	0,33	34,87	0,11	31,08
6	7,00	6,45	0,55	19,31	0,31	14,76
6	7,23	6,45	0,79	19,31	0,62	13,02
6	6,30	6,45	-0,15	19,31	0,02	20,63

FIGURA 40 - Adjustment of the growth equation for dominant height (h_{100}).

FIGURA 41 - Column calculations in Figura 40 for fitting the Chapman-Richards model for the dominant height $(h\ )_{100}$

Colum n	Label	Equation	Formula in 2nd line
M	X	= Age	=A2
N	Y	$= h_{100}$	=B2
O	Ye	$=A*(1-e(-k*M2))^r$	=$H$8*(1-EXP(-$H$9*M2))^$H$10
P	Waste	Y-Ye	=N2-O2
Q	Waste²	$(Ye-Ym)^2$	=(O2-$H$4)^2
R	Regression²	$(Y-Ye)^2$	=(N2-O2)^2
S	Total²	$(Y-Ym)^2$	=(N2-$H$4)^2

Where: The calculations must be repeated up to the last line of dominant heights.

The values for columns M to S in Excel are determined as follows Figura 41.

TABELA 21 - Cell calculations for analysis of variance of the regression of Figura 40 for fitting the equation for $(h\ )_{100}$

Cell	Label	Equation	Formula in 2nd line
H4	Ym	$\sum Y / n$	=AVERAGE(N2:N161)
H6	n	No. of observations	=CONT.NÚM(N2:N161)
H11	p	No. of parameters	=CONT.NÚM(H8:H10)
G15	GLreg	=p-1	=H11-1
G16	GLres	=n-p	=H6-H11
G17	GLtot	=n-1	=H6-1
H15	SQreg	$(Ye-Ym)^2$	=SUM(Q2:Q161)
H16	SQres	$(Y-Ye)^2$	=SUM(R2:R161)
H17	SQtot	$(Y-Ym)^2$	=SUM(S2:S161)
I15	QMreg	SQreg/GLreg	=H15/G15
I16	QMres	SQres/GLres	=H16/G16
I17	QMtot	SQtot/GLtot	=H17/G17
J15	F	QMreg/QMres	=I15/I16
K15	PR>F	Probab. From F	=DIST.F(J15;G15;G16;FALSE)
G19	Syx	$\sqrt{QMres}$	=RAIZ(I16)

Cell	Label	Equation	Formula in 2nd line
G20	CV	100.Syx/Ym	=100*G19/H4
G21	R^2	1-SQres/SQtot	=1-H16/H17
G22	R^2aj	R^2-(GLreg/GLres).(1-R^2)	=G21-(G15/G16)*(1-G21)

Where: Ym= mean of Y; n= number of observations; p= number of parameters in the equation; GLreg= degrees of freedom of the regression; GLres= degrees of freedom of the residual; GLtot= degrees of freedom of the total; SQreg= sum of squares of the regression; SQres= sum of squares of the residual; SQtot= sum of squares of the total; QMreg= mean square of the regression; QMres= mean square of the residual; QMtot= ; F= Snedecor's F value; PR>F= probability of the F value; Syx= standard error of estimate; CV= coefficient of variation; R^2= coefficient of determination; R^2aj= adjusted coefficient of determination.

Once the sums of squares (SQ) have been determined, EXCEL's SOLVER is used to adjust the equation, minimizing the Sum of Squares of the Residuals (SQ) of cell $H$16, as shown in Figure 42. Figura 42.

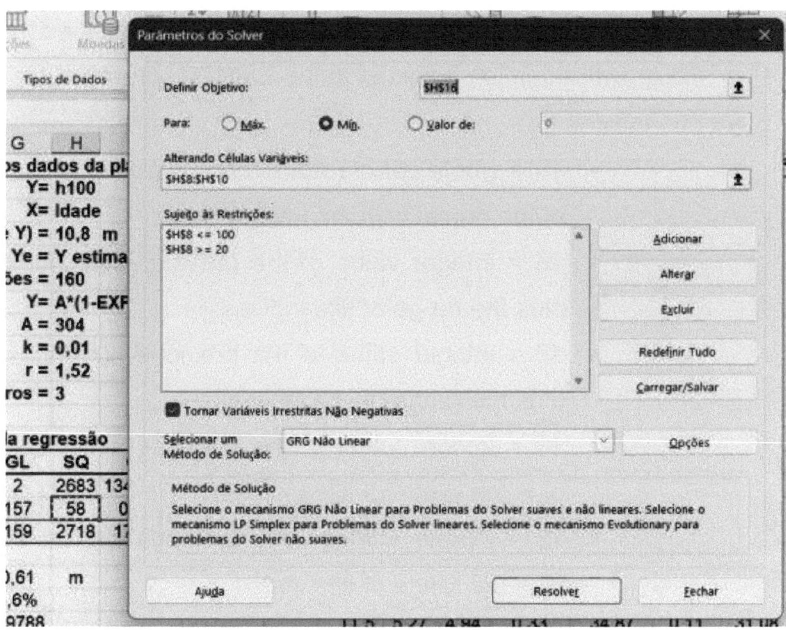

FIGURA 42 - Adjustment of the equation for estimates of h_{100} as a function of age.

6.6.4.6 Spreadsheet: Sites

Construction of the table and graph of site indices with the master curve of h_{100} estimated by the adjusted equation.

Using site indices with an interval of 1 meter and establishing the rotation age for the forest, the reference age is calculated at 70% of the rotation age, using the integer value of the h_{100} reached at that age, which will be the reference value for constructing the site index table. The range of site indices will be from the integer value rounded down from the h_{100} of the plot with the lowest value to the integer value rounded up from the h_{100} of the plot with the highest value.

The site index table will contain the following columns:

- Age - years of age of the forest;
- h100 - dominant height at the considered age calculated by the growth equation adjusted in the Eqh100 spreadsheet;
- Site Indices (m):
    - 15 = whole value rounded down from the h_{100} of the parcel with the lowest value;
    - 16 = integer value of the previous site index plus the range of site indices;
    - 17 = integer value of the previous site index plus the range of site indices;
    - 18 = integer value of the previous site index plus the range of site indices;
    - 19 = integer value of the previous site index plus the range of site indices;
    - 20 = integer value of the previous site index plus the range of site indices.

The calculations for the site index table (Figura 43) are made using the formulas in Tabela 22 below.

TABELA 22 - Calculations used to determine site indices

Cell	Label	Equation	Formula
E2	Age of rotation:	Established by technical criteria	18
E3	Reference age:	70% of rotation age	=INT(70%*E2)
E4	minimum h100 at the reference age:	Minimum h_{100}	=MÍNIMO(hdom!D142:D161)
E5	maximum h100 at the reference age:	Maximum h_{100}	=MÁXIMO(hdom!D142:D161)
E6	Range of site indices (smaller):	Lower IS	=ROUND.DOWN(E4;0)
G6	Range of site indices (greater):	Higher IS	=ROUND.UP(E5;0)
E7	Range of site indices:	Convenience	1
A14 to A29	Ages, from the youngest in the inventory to the age of rotation	According to inventory	5 a 18
B14 to B29	h_{100}	Master curve	=$E$9*(1- EXP(-$E$10*A14))^$E$11 a =$E$9*(1- EXP(-$E$10*A29))^$E$11
C13 to H13	Site Indices	IS from lowest to highest value	15 a 20
E13	Whole value of h_{100} at reference age	IS 17	=INT(B21)

Cell	Label	Equation	Formula
E21	Reference value	IS value at reference age	=E13
C121 to H21	h_{100} from smallest to largest at reference age	h_{100} at the reference age	= C13, D13, E13, F13, G13 and H13 respectively
E20	h_{100} at age younger than the reference age in the IS 17	h_{100} at 11	=E21*$B20/$B21
C14 to H20	h_{100} at ages younger than the reference age	Copy E20 and paste formula	=C15*$B14/$B15 to =H21*$B20/$B21
E22	h_{100} at age greater than the reference age in the IS 17	h_{100} at 13	=E21*$B22/$B21
C22 to H29	h_{100} at ages older than the reference age	Copy E22 and paste formula	=C21*$B22/$B21 to =H28*$B29/$B28

The site index table is shown in Figura 43 below.

6.6.4.7 Spreadsheet: IS

Table showing the site index (SI) of each plot and the average SI per site (stratum). To classify the SI from meter to meter, it is sufficient to convert the h_{100} of each plot into an integer value and for the site SI, the average of the h_{100} of the plots on the site must be taken and converted into an integer value.

E21 =E13

Índices de Sítio

Idade de rotação:	18 anos
Idade de referência:	12 anos
h_{100} mínima na idade de referência:	15,767 m
h_{100} máxima na idade de referência:	19,567 m
Amplitude de índices de sítio:	15 até 20 m
Intervalo de índices de sítio:	1 m
Equação de crescimento:	$h_{100} = A\,(1 - EXP(-k \cdot Idade))^r$
A =	303,70
k =	0,0138
r =	1,5212

Idade (anos)	h_{100} (m)	Índices de Sítio (m) 15	16	17	18	19	20
5	4,94	4,3	4,5	4,8	5,1	5,4	5,7
6	6,45	5,6	5,9	6,3	6,7	7,0	7,4
7	8,07	7,0	7,4	7,9	8,3	8,8	9,3
8	9,78	8,4	9,0	9,6	10,1	10,7	11,2
9	11,58	10,0	10,7	11,3	12,0	12,6	13,3
10	13,46	11,6	12,4	13,1	13,9	14,7	15,5
11	15,40	13,3	14,2	15,0	15,9	16,8	17,7
12	**17,40**	**15,0**	**16,0**	17,0	**18,0**	**19,0**	**20,0**
13	19,45	16,8	17,9	19,0	20,1	21,2	22,4
14	21,55	18,6	19,8	21,1	22,3	23,5	24,8
15	23,70	20,4	21,8	23,2	24,5	25,9	27,2
16	25,88	22,3	23,8	25,3	26,8	28,3	29,7
17	28,09	24,2	25,8	27,4	29,1	30,7	32,3
18	30,34	26,2	27,9	29,6	31,4	33,1	34,9
19	32,61	28,1	30,0	31,9	33,7	35,6	37,5
20	34,90	30,1	32,1	34,1	36,1	38,1	40,1

FIGURA 43 - Site Index Table.

In Figura 44 shows the results obtained in the classification of sites by plot and by location.

6.6.4.8 Spreadsheet: df

Adjustment of the equations for determining future diameter (df) as a function of diameter measurements from 5 to 12 years of age. The spreadsheet for adjusting the equation to estimate future diameter is shown in Figura 45. The equation is adjusted using EXCEL's SOLVER tool as shown in Figura 46. SOLVER is accessed from the Data menu.

E2		=INT(MÉDIA(C2:C5))		
Local	**Parcela**	**h_{100}**	**IS**	**IS do Local**
1	1	19,1	19	18,0
1	2	19,6	19	
1	3	17,1	17	
1	4	17,1	17	
2	1	16,1	16	16
2	2	16,5	16	
2	3	16,8	16	
2	4	17,0	17	
3	1	15,8	15	16
3	2	16,7	16	
3	3	16,7	16	
3	4	16,6	16	
4	1	17,1	17	17
4	2	17,2	17	
4	3	17,4	17	
4	4	17,8	17	
5	1	17,5	17	17
5	2	17,5	17	
5	3	18,0	17	
5	4	18,3	18	

FIGURA 44 - Site indices per plot and per location

6.6.4.9 Spreadsheet: hf

Adjustment of the equations for determining future height (hf) as a function of height measurements from 5 to 12 years. The spreadsheet for adjusting the equation to estimate future height is shown in Figura 47Figura 45. The equation is adjusted using EXCEL's SOLVER tool, as shown in Figura 48. SOLVER is accessed from the Data menu.

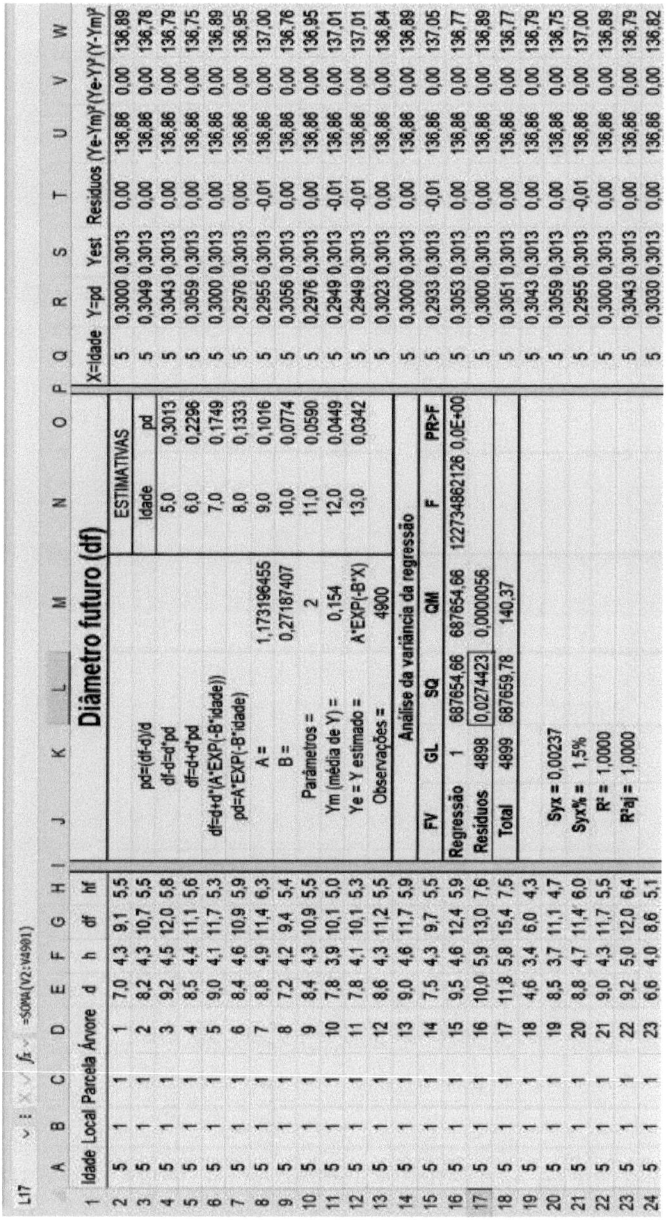

L17 =SOMA(V2:V4901)

Idade	Local	Parcela	Árvore	d	h	df	hf
5	1	1	1	7,0	4,3	9,1	5,5
5	1	1	2	8,2	4,3	10,7	5,5
5	1	1	3	9,2	4,5	12,0	5,8
5	1	1	4	8,5	4,4	11,1	5,6
5	1	1	5	9,0	4,1	11,7	5,3
5	1	1	6	8,4	4,6	10,9	5,9
5	1	1	7	8,8	4,9	11,4	6,3
5	1	1	8	7,2	4,2	9,4	5,4
5	1	1	9	8,4	4,3	10,9	5,5
5	1	1	10	7,8	3,9	10,1	5,0
5	1	1	11	7,8	4,1	10,1	5,3
5	1	1	12	8,6	4,3	11,2	5,5
5	1	1	13	9,0	4,6	11,7	5,9
5	1	1	14	7,5	4,3	9,7	5,5
5	1	1	15	9,5	4,6	12,4	5,9
5	1	1	16	10,0	5,9	13,0	7,6
5	1	1	17	11,8	5,8	15,4	7,5
5	1	1	18	4,6	3,4	6,0	4,3
5	1	1	19	8,5	3,7	11,1	4,7
5	1	1	20	8,8	4,7	11,4	6,0
5	1	1	21	9,0	4,3	11,7	5,5
5	1	1	22	9,2	5,0	12,0	6,4
5	1	1	23	6,6	4,0	8,6	5,1

Diâmetro futuro (df)

pd=(df-d)/d

df-d=d*pd

df=d+d*pd

df=d+d*(A*EXP(-B*idade))

pd=A*EXP(-B*idade)

A =	1,173196455
B =	0,27187407
Parâmetros =	2
Ym (média de Y) =	0,154
Ye = Y estimado =	A*EXP(-B*X)
Observações =	4900

ESTIMATIVAS

Idade	pd
5,0	0,3013
6,0	0,2296
7,0	0,1749
8,0	0,1333
9,0	0,1016
10,0	0,0774
11,0	0,0590
12,0	0,0449
13,0	0,0342

Análise da variância da regressão

FV	GL	SQ	QM	F	PR>F
Regressão	1	687654,66	687654,66	122734862126	0,0E+00
Resíduos	4898	0,0274423	0,0000056		
Total	4899	687659,78	140,37		

Syx = 0,00237

Syx% = 1,5%

R² = 1,0000

R²aj = 1,0000

X=Idade	Y=pd	Yest	Resíduos	(Ye-Ym)²	(Ye-Y)²	(Y-Ym)²
5	0,3000	0,3013	0,00	136,86	0,00	136,89
5	0,3049	0,3013	0,00	136,86	0,00	136,78
5	0,3043	0,3013	0,00	136,86	0,00	136,79
5	0,3059	0,3013	0,00	136,86	0,00	136,75
5	0,3000	0,3013	0,00	136,86	0,00	136,89
5	0,2976	0,3013	0,00	136,86	0,00	136,95
5	0,2955	0,3013	-0,01	136,86	0,00	137,00
5	0,3056	0,3013	0,00	136,86	0,00	136,76
5	0,2976	0,3013	0,00	136,86	0,00	136,95
5	0,2949	0,3013	-0,01	136,86	0,00	137,01
5	0,2949	0,3013	-0,01	136,86	0,00	137,01
5	0,3023	0,3013	0,00	136,86	0,00	136,84
5	0,3000	0,3013	0,00	136,86	0,00	136,89
5	0,2933	0,3013	-0,01	136,86	0,00	137,05
5	0,3053	0,3013	0,00	136,86	0,00	136,77
5	0,3000	0,3013	0,00	136,86	0,00	136,89
5	0,3051	0,3013	0,00	136,86	0,00	136,77
5	0,3043	0,3013	0,00	136,86	0,00	136,79
5	0,3059	0,3013	0,00	136,86	0,00	136,75
5	0,2955	0,3013	-0,01	136,86	0,00	137,00
5	0,3000	0,3013	0,00	136,86	0,00	136,89
5	0,3043	0,3013	0,00	136,86	0,00	136,79
5	0,3030	0,3013	0,00	136,86	0,00	136,82

FIGURA 45 - Adjustment of the equation for future diameter (df).

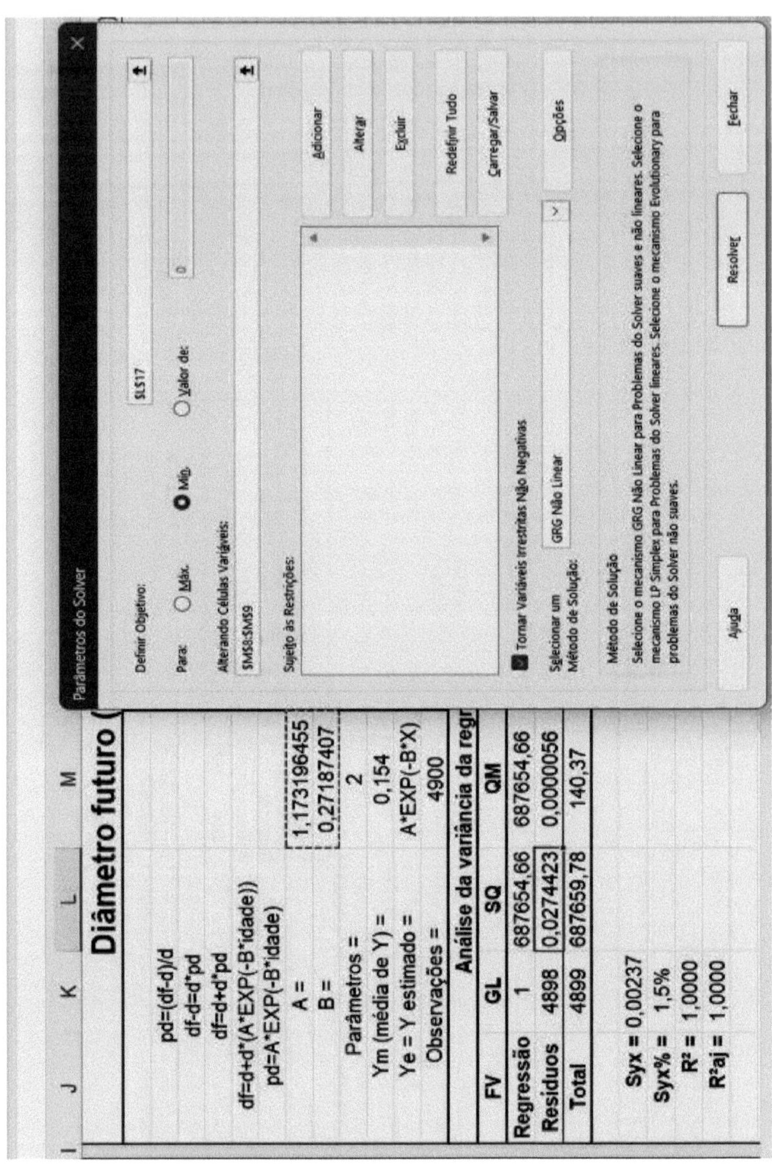

FIGURA 46 - Adjusting the equation for df with SOLVER.

L17 =SOMA(V2:V4901)

	Idade	Local	Parcela	Árvore	d	h	df	hf
2	5	1	1	1	7,0	4,3	9,1	5,5
3	5	1	1	2	8,2	4,3	10,7	5,5
4	5	1	1	3	9,2	4,5	12,0	5,8
5	5	1	1	4	8,5	4,4	11,1	5,6
6	5	1	1	5	9,0	4,1	11,7	5,3
7	5	1	1	6	8,4	4,6	10,9	5,9
8	5	1	1	7	8,8	4,9	11,4	6,3
9	5	1	1	8	7,2	4,2	9,4	5,4
10	5	1	1	9	8,4	4,3	10,9	5,5
11	5	1	1	10	7,8	3,9	10,1	5,0
12	5	1	1	11	7,8	4,1	10,1	5,3
13	5	1	1	12	8,6	4,3	11,2	5,5
14	5	1	1	13	9,0	4,6	11,7	5,9
15	5	1	1	14	7,5	4,3	9,7	5,5
16	5	1	1	15	9,5	4,6	12,4	5,9
17	5	1	1	16	10,0	5,9	13,0	7,6
18	5	1	1	17	11,8	5,8	15,4	7,5
19	5	1	1	18	4,6	3,4	6,0	4,3
20	5	1	1	19	8,5	3,7	11,1	4,7
21	5	1	1	20	8,8	4,7	11,4	6,0
22	5	1	1	21	9,0	4,3	11,7	5,5
23	5	1	1	22	9,2	5,0	12,0	6,4
24	5	1	1	23	6,6	4,0	8,6	5,1

Altura futura (df)

		ESTIMATIVAS	
		Idade	pd
ph=(hf-h)/h		5,0	0,2837
hf-h=h*ph		6,0	0,2477
hf=h+h*ph		7,0	0,2162
hf=h+h*(A*EXP(-B*idade))		8,0	0,1887
ph=A*EXP(-B*idade)		9,0	0,1647
A =	0,560066	10,0	0,1437
B =	0,136001	11,0	0,1255
Parâmetros =	2	12,0	0,1095
Ym (média de Y) =	0,196	13,0	0,0956
Ye (Y estimado)			
Observações =	4900		

Análise da variância da regressão

FV	GL	SQ	QM	F	PR>F
Regressão	1	682782,98	682782,98	45215988107,7	0,0E+00
Resíduos	4898	0,0739621	0,0000151		
Total	4899	682785,17	139,37		

Syx = 0,0039

Syx% = 2,0%

R² = 1,0000

R²aj = 1,0000

X=Idade	Y=ph	Yest	Resíduos	(Ye-Ym)²	(Ye-Y)²	(Y-Ym)²
5	0,2791	0,2837	0,00	137,27	0,00	137,38
5	0,2791	0,2837	0,00	137,27	0,00	137,38
5	0,2889	0,2837	0,01	137,27	0,00	137,15
5	0,2727	0,2837	-0,01	137,27	0,00	137,53
5	0,2927	0,2837	0,01	137,27	0,00	137,06
5	0,2826	0,2837	0,00	137,27	0,00	137,30
5	0,2857	0,2837	0,00	137,27	0,00	137,22
5	0,2857	0,2837	0,00	137,27	0,00	137,22
5	0,2791	0,2837	0,00	137,27	0,00	137,38
5	0,2821	0,2837	0,00	137,27	0,00	137,31
5	0,2927	0,2837	0,01	137,27	0,00	137,06
5	0,2791	0,2837	0,00	137,27	0,00	137,38
5	0,2826	0,2837	0,00	137,27	0,00	137,30
5	0,2791	0,2837	0,00	137,27	0,00	137,38
5	0,2826	0,2837	0,00	137,27	0,00	137,30
5	0,2881	0,2837	0,00	137,27	0,00	137,17
5	0,2931	0,2837	0,01	137,27	0,00	137,05
5	0,2647	0,2837	-0,02	137,27	0,00	137,72
5	0,2703	0,2837	-0,01	137,27	0,00	137,59
5	0,2766	0,2837	-0,01	137,27	0,00	137,44
5	0,2791	0,2837	0,00	137,27	0,00	137,38
5	0,2800	0,2837	0,00	137,27	0,00	137,36
5	0,2750	0,2837	-0,01	137,27	0,00	137,48

FIGURA 47 - Adjustment of the equation for future height (hf).

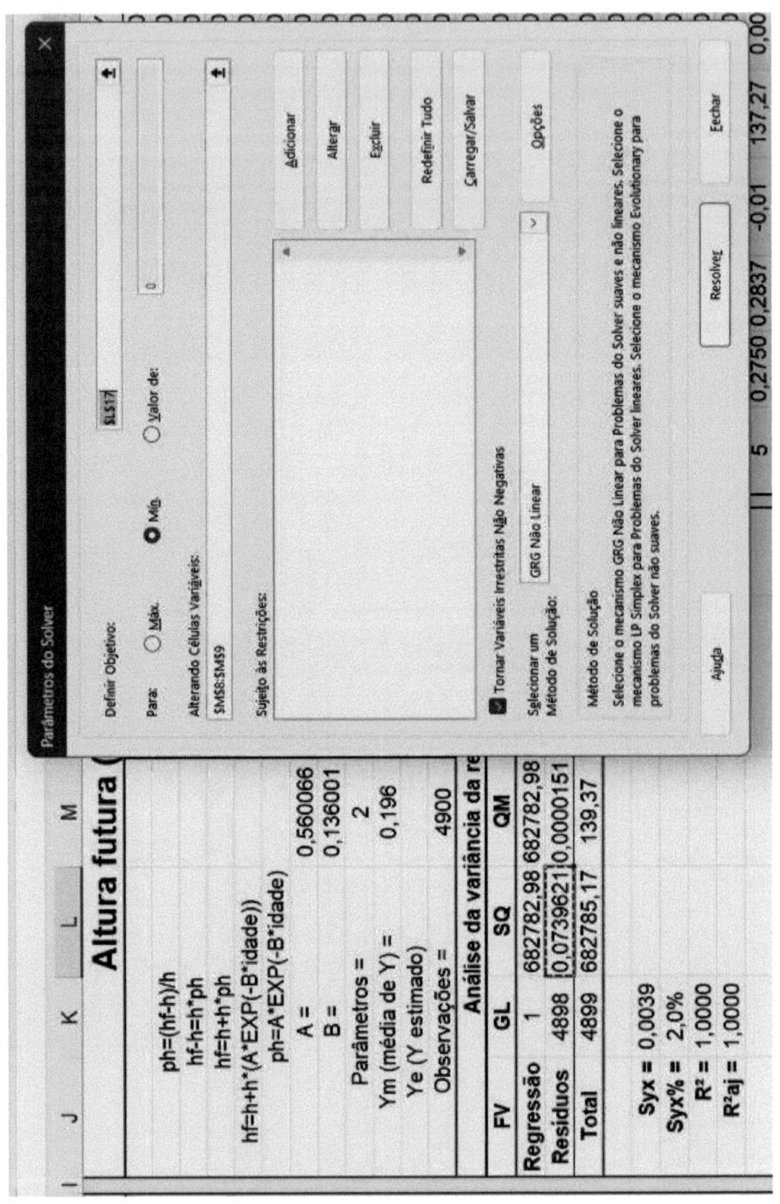

FIGURA 48 - Adjusting the equation for hf with SOLVER.

6.6.4.10 Spreadsheet: 13 years

This table shows, as an example, the estimates of tree diameter and height in the plots not measured at 13 years (Tabela 23).

TABELA 23 - Estimates of diameter and height of plot 1 at site 1.

Age	Location	Tranche	Tree	d	h	df (13years)	hf (13years)
12	1	1	1	18,8	14,9	19,4	16,3
12	1	1	2	22,1	14,9	22,9	16,3
12	1	1	3	24,8	15,6	25,6	17,1
12	1	1	4	22,7	15,2	23,5	16,7
12	1	1	5	24,1	14,3	24,9	15,7
12	1	1	6	22,3	16,1	23,1	17,6
12	1	1	7	23,5	17,1	24,3	18,7
12	1	1	8	19,2	14,4	19,9	15,8
12	1	1	9	22,3	14,9	23,1	16,3
12	1	1	10	20,7	13,3	21,4	14,6
12	1	1	11	20,7	14,3	21,4	15,7
12	1	1	12	23,2	14,9	24,0	16,3
12	1	1	13	24,1	16,1	24,9	17,6
12	1	1	14	19,9	14,9	20,6	16,3
12	1	1	15	25,6	16,1	26,5	17,6
12	1	1	16	26,8	20,7	27,7	22,7
12	1	1	17	31,9	20,4	33,0	22,3
12	1	1	18	12,3	11,2	12,7	12,3
12	1	1	19	22,7	12,7	23,5	13,9
12	1	1	20	23,5	16,2	24,3	17,7
12	1	1	21	24,1	14,9	24,9	16,3
12	1	1	22	24,8	17,2	25,6	18,8
12	1	1	23	17,6	13,8	18,2	15,1
12	1	1	24	24,8	16,2	25,6	17,7
12	1	1	25	21,3	17,1	22,0	18,7
12	1	1	26	21,3	14,9	22,0	16,3
12	1	1	27	18,8	14,9	19,4	16,3
12	1	1	28	16,1	12,7	16,7	13,9
12	1	1	29	24,1	14,4	24,9	15,8
12	1	1	30	26,4	16,1	27,3	17,6
12	1	1	31	21,3	15,2	22,0	16,7
12	1	1	32	24,1	16,1	24,9	17,6
12	1	1	33	25,9	16,1	26,8	17,6
12	1	1	34	23,5	17,1	24,3	18,7
12	1	1	35	22,1	14,9	22,9	16,3

Where: d = diameter at 1.3 m height; h = total height; df = future diameter; hf = future height.

The diameters (df) at 13 years of age were estimated using the adjusted equation (Figura 45) expressed below:

$$df = d + d\ .\ (1,173196455\ .\ e^{(-\ 0,27187407\ .\ t)})$$

Where: df = future diameter; d = current diameter; e = base of the natural logarithm (2.718282); t = age in years.

The heights (hf) at 13 years of age were estimated using the adjusted equation (Figura 47) expressed below:

$$hf = h + h\ .\ (0.560066174\ .\ e\ )^{(-\ 0,136001\ .\ t)}$$

Where: hf = future height; d = current height; e = base of the natural logarithm (2.718282); t = age in years.

6.6.4.11 Spreadsheet: Cubage

The cubing data for 24 trees can be found in Appendix A, with the following variables:

- Tree - number of the tree felled;
- Section - number of the tree section;
- d - diameter of the tree at 1.3 m from the ground;
- e - simple bark thickness at 1.3 m from the ground;
- h - total height of the tree;
- d_i - trunk diameter at position i;
- and_i - simple shell thickness at position i;
- $2and_i$ - bark thickness multiplied by 2;
- $Comp_i$ - length of section i;
- S - type of log assortment according to its diameter at the thin end (t = stump; s1 = log (<20cm); s2 = thin log (≥20-<25cm); s3 = medium log (≥25-<30cm; s4 = thick log (≥30cm));
- h_i - height of the tree in position i;
- V_i - shell volume of section i calculated using the Smalian method;

- d_{si} - diameter without shell at position i = $d - 2e_i$;
- V_{si} - volume without shell of section i calculated by the Smalian method.

6.6.4.12 Spreadsheet: Volume

The Volume spreadsheet shows the results of the tree volume calculation (Tabela 24) and the Spurr model fit of the combined variable ($v=b0+b1.d^2.h$) for the total volume of trees with bark.

TABELA 24 - Diameter, height, assortment volume and total volume of the 24 trees cubed

Tree	d	e	h	t	m	s1	s2	s3	s4	Waste	v
1	29,0	1,0	27,5	0,01281	0,03468	0,11630	0,32228	0,46545	0,00000	0,05398	1,00549
2	32,5	2,2	25,5	0,01248	0,05286	0,11794	0,21748	0,29737	0,22035	0,01702	0,93550
3	38,0	2,6	28,0	0,03324	0,03711	0,06763	0,11075	0,48584	0,78069	0,04599	1,56126
4	22,5	0,9	24,5	0,01260	0,08623	0,22232	0,09503	0,11310	0,00000	0,01740	0,54668
5	37,0	1,4	28,7	0,02694	0,03468	0,06597	0,10843	0,47409	0,98996	0,02788	1,72795
6	18,5	1,2	18,7	0,00424	0,09296	0,17113	0,00000	0,00000	0,00000	0,03801	0,30634
7	34,0	2,0	26,3	0,01977	0,07344	0,07274	0,11310	0,47658	0,46121	0,02566	1,24248
8	39,0	2,5	27,0	0,03064	0,03588	0,07274	0,14050	0,17378	1,22027	0,03845	1,71227
9	38,5	2,0	27,1	0,01963	0,03711	0,06931	0,11786	0,33114	0,98700	0,03996	1,60202
10	33,0	1,8	22,7	0,01162	0,06309	0,00000	0,20835	0,70303	0,00000	0,01139	0,99748
11	33,5	2,0	20,3	0,01735	0,07950	0,00000	0,12768	0,32792	0,48526	0,01173	1,04943

Tree	d	e	h	t	m	s1	s2	s3	s4	Waste	v
12	25,5	1,4	20,3	0,01078	0,02577	0,10878	0,00000	0,42456	0,00000	0,01843	0,58833
13	28,0	1,0	24,1	0,00931	0,04627	0,06597	0,21267	0,46610	0,00000	0,08036	0,88070
14	28,0	1,3	22,2	0,00916	0,03836	0,05954	0,21050	0,45968	0,00000	0,04778	0,82502
15	23,0	1,3	20,8	0,00755	0,05345	0,10582	0,28377	0,00000	0,00000	0,02036	0,47094
16	22,5	2,0	22,8	0,00493	0,05865	0,16683	0,19764	0,00000	0,00000	0,03770	0,46575
17	21,5	1,0	22,1	0,00664	0,06311	0,10565	0,28412	0,00000	0,00000	0,03521	0,49473
18	16,0	0,9	18,1	0,00318	0,08979	0,05245	0,00000	0,00000	0,00000	0,02513	0,17055
19	17,0	0,9	19,1	0,00277	0,06661	0,13017	0,00000	0,00000	0,00000	0,02714	0,22670
20	17,0	1,0	21,4	0,00407	0,10834	0,14117	0,00000	0,00000	0,00000	0,03456	0,28814
21	13,0	1,0	13,1	0,00229	0,07219	0,00000	0,00000	0,00000	0,00000	0,02842	0,10290
22	14,0	0,7	18,7	0,00291	0,13026	0,00000	0,00000	0,00000	0,00000	0,03435	0,16753
23	14,0	0,9	16,6	0,00153	0,06982	0,04048	0,00000	0,00000	0,00000	0,03828	0,15010
24	13,0	0,5	17,7	0,00215	0,09234	0,00000	0,00000	0,00000	0,00000	0,03480	0,12929

Where: Tree = number of the felled tree; d = diameter of the tree at 1.3 m from the ground; e = simple thickness of the bark at 1.3 m from the ground; h = total height of the tree; t = stump; s1 = log (<20cm); s2 = thin log (≥20-<25cm); s3 = medium log (≥25-<30cm; s4 = thick log (≥30cm); Residue = non-commercial part of the trunk; v = total tree volume (m^3).

The Spurr volumetric model ($v=b0+b1.d^2.h$) was adjusted using EXCEL's Data menu, with the Regression tool (Figura 49), with the tree volume as Y and the combined variable $d^2.h$ as X, resulting in the equation:

$$v = 0.034046 + 0.0000402 \,.\, d^2.h$$

Where: v = total volume estimated for the tree considered; d = diameter of the tree at 1.3 m from the base in centimeters; h = total height of the tree in meters.

y=volume	x=d²h	RESUMO DOS RESULTADOS								
1,00549	23135,9									
0,93550	26944,9	*Estatística de regressão*								
1,56126	40432,0	R múltiplo	0,99266							
0,54668	12413,3	R-Quadrado	0,98537							
1,72795	39345,1	R-quadrado aju	0,98470							
0,30634	6386,4	Erro padrão	0,06621	m³.	CV =	9,00%				
1,24248	30437,5	Observações	24							
1,71227	40991,0	V médio	0,73532							
1,60202	40169,0	ANOVA								
0,99748	24698,5		*gl*	*SQ*	*MQ*	*F*	*F de significação*			
1,04943	22781,7	Regressão	1	6,4949	6,4949	1481,41	1,1157E-21			
0,58833	13219,6	Resíduo	22	0,0965	0,0044					
0,88070	18878,7	Total	23	6,5913						
0,82502	17397,0									
0,47094	11003,2		*Coeficientes*	*Erro padrão*	*Stat t*	*valor-P*	*95% inferiores*	*95% superiores*	*Inferior 95.0%*	*Superior 95.0%*
0,46575	11532,4	Interseção	0,034046	0,022686	1,500759	0,147634	-0,013002	0,081093	-0,013002	0,081093
0,49473	10206,5	Variável X 1	4,02E-05	1,05E-06	3,85E+01	1,12E-21	3,81E-05	4,24E-05	3,81E-05	4,24E-05
0,17055	4628,5									
0,22670	5514,1									
0,28814	6181,7									
0,10290	2212,2									
0,16753	3659,3									
0,15010	3245,8									
0,12929	2987,9									

FIGURA 49 - Adjustment of the Spurr model for total tree volume.

6.6.4.13 Spreadsheet: V (m³/ha)

It shows a table with the bark volumes of the trees estimated by the adjusted equation (Tabela 25), at 13 years of age, with the following variables (spreadsheet columns):

- Age - age of the trees;
- Location - inventory layer;
- Plot - plot number in the stratum;
- Tree - number of the tree in the plot;
- d - diameter of the tree at a height of 1.3 m;
- h - total height of the tree;
- v - total individual volume with tree bark, estimated by the adjusted equation ($v = 0.034046 + 0.0000402 . d^2.h$) , in m^3;
- Vi - volume with bark per hectare that tree i represents in the plot in m^3/ha, calculated by Vi = v . 10000 / plot area ;
- Viac - volume with bark per hectare in m^3/ha accumulated tree by tree in the plot;
- V - volume with bark per hectare of the plot in m^3/ha.

The formulas used to calculate the individual volumes per tree and per hectare were:

- $v = 0.034046 + 0.0000402 . d^2.h$;
- Vi = v . 10000 / 294;
- Viac = V(i-1)ac + Vi, where V(i-1)ac = volume per hectare accumulated up to tree order i-1;
- V (m^3/ha) = SE(Tree=35;Viac;"") - if the tree number is equal to 35, which is the last tree in the plot, then the accumulated volume up to it is the volume per hectare of the plot.

TABELA 25 - Individual tree volume and per hectare of the plots

Age	Location	Tranche	Tree	d	h	v	I saw	Viac	V (m³/ha)
13	1	1	1	19,4	16,3	0,282297	9,60	9,60	
13	1	1	2	22,9	16,3	0,377098	12,83	22,43	
13	1	1	3	25,6	17,1	0,486336	16,54	38,97	
					...				
13	1	1	33	26,8	17,6	0,543159	18,47	471,27	
13	1	1	34	24,3	18,7	0,479211	16,30	487,57	
13	1	1	35	22,9	16,3	0,377098	12,83	500,39	**500,39**
13	1	2	1	19,5	14,9	0,261955	8,91	8,91	
13	1	2	2	26,8	18,1	0,556990	18,95	27,86	
13	1	2	3	17,9	13,3	0,205467	6,99	34,84	
					...				
13	5	4	33	38,2	18,1	1,096505	37,30	831,44	
13	5	4	34	25,9	16,2	0,471187	16,03	847,47	
13	5	4	35	37,2	18,1	1,041607	35,43	882,90	**882,90**

6.6.4.14 Spreadsheet: datap

The volume per hectare data from the pilot inventory plots is shown in Tabela 26.

TABELA 26 - Volumes per hectare of the plots in the pilot inventory at 13 years of age

Age	Location	Tranche	V (m^3/ha)
13	1	1	500,39
13	1	2	434,63
13	1	3	584,90
13	1	4	596,36
13	2	1	328,87
13	2	2	308,88
13	2	3	444,53
13	2	4	522,36
13	3	1	297,44
13	3	2	287,87
13	3	3	405,18
13	3	4	412,47
13	4	1	569,74
13	4	2	541,58
13	4	3	900,03
13	4	4	807,93
13	5	1	650,31
13	5	2	597,61
13	5	3	859,05
13	5	4	882,90

6.6.4.15 Worksheet: Pilot

The results of the pilot inventory data processing are shown in Tabela 27 and in Figura 50 shows the process of determining sampling sufficiency.

TABELA 27 - Results of the pilot inventory

Age	Location	Tranche	V (m^3/ha)	$\overline{x}_h$	Sh^2	nh	Area (ha)	Ph	$\overline{\overline{x}}$	S^2	$Sh^2.Ph / S^2$
13	1	1	500,4	529,07	5795,53	4	17,00	0,17	547,9	14295,07	0,06892
13	1	2	434,6								
13	1	3	584,9								
13	1	4	596,4								

Age	Location	Tranche	V (m³/ha)	$\overline{x}_h$	Sh²	n h	Area (ha)	Ph	$\overline{\overline{x}}$	S²	Sh².Ph / S²
13	2	1	328,9	401,16	10103,42	4	28,00	0,28			0,19790
13	2	2	308,9								
13	2	3	444,5								
13	2	4	522,4								
13	3	1	297,4	350,74	4522,77	4	15,00	0,15			0,04746
13	3	2	287,9								
13	3	3	405,2								
13	3	4	412,5								
13	4	1	569,7	704,82	31210,57	4	14,00	0,14			0,30566
13	4	2	541,6								
13	4	3	900,0								
13	4	4	807,9								
13	5	1	650,3	747,47	20896,05	4	26,00	0,26			0,38006
13	5	2	597,6								
13	5	3	859,0								
13	5	4	882,9								
Total	5	20					100,00				1,00000

Where: $\overline{\overline{x}}$ = stratified mean of the population; $\overline{x}_h$ = mean of stratum h; P_h = proportion of stratum h over the total area; n = number of strata; h = order number of the 1st level unit; S_h^2 = variance of stratum h; S^2= stratified variance of the population; i = order number of the sampling unit in stratum h; N = potential number of sampling units in the population.

Suficiência Amostral						
Área populacional =					100,0	ha
Área da UA =					294	m²
N potencial da população =					**3401**	unidades
Probabilidade de confiança para a média					95%	
Média =					547,9	m³/ha
Erro amostral relativo =					10,0%	
Erro amostral absoluto =					**54,8**	m³/ha
Variância =					**14295**	(m³/ha)²
n amostral =					20	UA
Graus de liberdade =		19	20	20		
t (prob; gl) =		2,093	2,086	2,086		
n suficiente =		20,99	20,85	20,85	**21**	**UAs**
Local (j)	nj	Parcelas necessárias	Piloto	Parcelas efetivas	Adicionar parcelas	
1	1,45	2	4	4	0	
2	4,16	5	4	5	1	
3	1,00	1	4	4	0	
4	6,42	7	4	7	3	
5	7,98	8	4	8	4	
Total				28	8	

São necessárias mais 1 parcela no local 2, 3 parcelas no Local 4 e 4 parcelas no local 5 para completar a amostragem no inventário definitivo, num total de 8 parcelas a acrescentar.

FIGURA 50 - Determining sample sufficiency.

6.6.4.16 Spreadsheet: DataD

The data from the definitive inventory plots is shown in Tabela 28with the volumes per hectare per sample plot.

TABELA 28 - Volumes per hectare of the plots in the definitive inventory

Age	Location	Tranche	V (m³/ha)
13	1	1	500,39
13	1	2	434,63
13	1	3	584,90
13	1	4	596,36
13	2	1	328,87
13	2	2	308,88
13	2	3	444,53
13	2	4	522,36
13	2	5	339,00
13	3	1	297,44
13	3	2	287,87

Age	Location	Tranche	V (m³/ha)
13	3	3	405,18
13	3	4	412,47
13	4	1	569,74
13	4	2	541,58
13	4	3	900,03
13	4	4	807,93
13	4	5	701,00
13	4	6	841,00
13	4	7	643,00
13	5	1	650,31
13	5	2	597,61
13	5	3	859,05
13	5	4	882,90
13	5	5	845,00
13	5	6	832,00
13	5	7	807,00
13	5	8	830,00

6.6.4.17 Spreadsheet: Definitive

The processing and results of the final inventory data are shown in Figura 51 e Figura 52.

Idade	Local	Parcela	V (m³/ha)	$\overline{x}_h$	Sh²	nh	Área (ha)	Ph	$\overline{\overline{x}}$	s²	Sh².Ph / S²	Entre	Dentro	Total	Ph² Sh² / nh	Ph Sh² / N
13	1	1	500,4	529,07	5795,53	4	17,00	0,17	547,00	8232,22	0,119681	1286,00	822,5	2172,5	41,87	0,2897
13	1	2	434,6										8919,0	12627,3		
13	1	3	584,9										3117,0	1436,4		
13	1	4	596,4										4528,0	2436,4		
13	2	1	328,9	355,32	3693,36	4	28,00	0,28			0,125621	146971,03	699,5	47581,6	72,39	0,3040
13	2	2	308,9										2156,5	56702,2		
13	2	3	444,5										7957,7	10501,7		
13	2	5	339,0										266,3	43265,7		
13	3	1	297,4	350,74	4522,77	4	15,00	0,15			0,082410	154080,10	2840,8	62282,4	25,44	0,1995
13	3	2	287,9										3952,9	67152,3		
13	3	3	405,2										2963,5	20115,0		
13	3	4	412,5										3811,1	18098,6		
13	4	1	569,7	714,90	19217,03	7	14,00	0,14			0,326811	197316,17	21071,0	516,8	53,81	0,7910
13	4	2	541,6										30038,4	29,4		
13	4	3	900,0										34272,8	124624,6		
13	4	4	807,9										8655,7	68083,9		
13	4	5	701,0										193,1	23714,8		
13	4	6	841,0										15902,0	86433,6		
13	4	7	643,0										5169,2	9215,2		
13	5	1	650,3	787,98	10938,62	8	26,00	0,26			0,345477	464566,01	18954,3	10671,8	92,43	0,8361
13	5	2	597,6										36241,7	2561,0		
13	5	3	859,0										5050,2	97371,2		
13	5	4	882,9										9008,5	112823,5		
13	5	5	845,0										3251,0	88801,6		
13	5	6	832,0										1937,5	81222,7		
13	5	7	807,0										361,7	67597,9		
13	5	8	830,0										1765,4	80086,7		
Totais	5	27					100,00				1,000000	964219,32	233907,5	1198126,8	285,94	2,4203

FIGURA 51 - Averages and variances in the final inventory

The variables and statistics in Figura 51 are as follows:

- $\bar{\bar{x}}$ = stratified average of the population;
- $\bar{x}_h$ = average of stratum of order h;
- P_h = proportion of the h-order stratum over the total area;
- N = number of strata;
- h = order number of the 1st level unit.
- $S_h{}^2$ = variance of stratum of order j;
- S^2 = stratified variance of the population;
- i = order number of the sampling unit in stratum h;
- N = potential number of sample units in the population.

The sampling intensity in the definitive inventory was 0.79% and the population was considered infinite.

A Figura 52 shows the results of the analysis of variance and the statistical results of the sampling in the final inventory.

Análise da variância

FV	GL	SQ	QM	F	PR>F
Entre estratos	4	964219	241055	22,67	6E-08
Dentro dos estratos	22	233908	10632	-	-
Total	26	1198127	-	-	-

** - O valor de F é altamente significativo, há diferença entre os estratos.

Coeficiente de variação CV = 18,9%

Variância da média $s^2_{\bar{x}}$ = 283,5 $(m^3/ha)^2$

Erro padrão da média $s_{\bar{x}}$ = 16,8 m^3/ha

Erro amostral E = 34,6 m^3/ha E% = 6,3 %

Intervalo de confiança IC = [512,4 m^3/ha ≤ média ≤ 581,6 m^3/ha] = 95%

Estimativa mínima de confiança

Para a média: EMC_m = 512,4 m^3/ha

Para o total: EMC_t = 51239 m^3

FIGURA 52 - Sampling results for the final inventory .

7. PLANNING FOREST INVENTORIES

Planning forest inventories involves nine stages, which are listed below:

1. Characterize the inventory
2. Carry out a preliminary inventory
3. Drawing up a plan
4. Training staff
5. Carry out a survey
6. Processing data
7. Analyzing results
8. Report
9. Presenting/disseminating results

The phases of each of the nine stages of forest inventory planning are described below.

7.1 Characterization of Forest Inventories

What characterizes forest inventories are: the object of the inventory (Forest and Environment); the objectives of the inventory; the precision required; the approach over time; and, the form of the survey (census or sampling process).

7.1.1 Object of Inventory

The object should be described in terms of the following components: the forest or vegetation, the objectives of the forest, the space occupied, the location and the environment.

7.1.1.1 Forest or vegetation

Characterize the type of forest. If it is natural, what is its successional state, what typologies are present in the inventory area. If it is planted, what species, planting date, initial density, management method, rotation age, etc.

7.1.1.2 Forest objectives

The objectives of the forest can be to produce wood for different purposes, or to produce non-timber products, or for conservation and preservation, or even for leisure.

7.1.1.3 Space

The area must have its boundaries defined and be mapped, measuring the total area, forest, preserved areas, infrastructure, water surfaces, etc.

The forest must be located with coordinates, identifying the municipality and state where it is located.

7.1.1.4 Environment

The description of the environment involves the natural environment with its topography, drainage network, climate, soils, etc. and the man-made environment, with reference to the owner, access, buildings.

7.1.2 Inventory objectives

A forest inventory can have different objectives, and this implies different approaches. The main objectives include:

- Public or private planning;

- Strategic, tactical or operational planning;
- Phytosociological;
- Ecological EIA, UC;
- Wood sales;
- Harvest planning;
- Forest management - stock, growth, sites.

7.1.3 Accuracy required

The accuracy required for a forest inventory depends on the criticality of the objective for which it is intended, the financial resources available, the relevant legislation or in agreement with the environmental agency involved, with the risks arising from low accuracy, among other factors.

When the accuracy required is 100%, a census is carried out, which is a complete enumeration of all the individuals in a population or community, obtaining the true parameters of the population, with no margin for error.

When the required precision is less than 100%, sampling is carried out, which is the measurement of a fraction of the population to represent it, obtaining statistics that estimate the true values (parameters) of the population with a certain probability of error.

Sampling can be done with random distribution of the sample units, in which case the emphasis is on estimating the variance, which is estimated with knowledge of the probability of error. Or, it can be carried out systematically, which implies greater precision of the mean, but the variance is estimated without knowing the error in estimating it.

The precision of a forest inventory is represented by the limit of the relative sampling error allowed around the mean, plus or minus (+/-E%) given as a percentage and by the probability of confidence in the mean (p).

The probability of confidence for the mean generally varies between 0.90 and 0.975 and the relative error limit (E%) is calculated as:

$$E\% = +/- \ 100 \ . \ 2 \ (1-p)$$

Where: E% = error limit around the mean in percent; p = probability of confidence for the mean.

A confidence probability for the mean of 0.975 is generally used for pre-cut inventories, when it is necessary to know more precisely the amount of wood available for sale, purchase or industrial supply. A probability of 0.95 is usually used for continuous inventories of planted forests, while a confidence probability of 0.9 can be used for monitoring managed native forests. In timber inventories for harvest planning in natural forest management, the probability is 1, i.e. 100%, since a census is required of the trees that can be harvested each year and, in this case, the mean is the true population mean resulting in the population parameter and there is no tolerance for error around the mean, the error is zero. When the probability of certainty for the mean is less than 1, the results are statistics that estimate the parameters; the statistics only approximate the true values and have a probability of error associated with the estimates.

Examples commonly used for confidence probability for the mean (P) and sampling error limit (LE%):

- Pre-cut inventory:

p = 0.975 and LE%=+/- 100 . 2 . (1-0,975) = +/- 5%;

- Continuous inventory of planted forests:

p = 0.950 and LE%=+/- 100 . 2 . (1-0,950) = +/- 10%;

- Continuous inventory of natural forests:

p = 0.900 and LE%=+/- 100 . 2 . (1-0,900) = +/- 20%.

7.1.4 Time approach

7.1.4.1 Temporary inventories

They are used to analyze and obtain data about the forest for the current situation. They are used for short-term planning, such as harvesting a particular forest stand.

Temporary forest inventories can be one-off, called Temporary Inventories, where temporary sampling units are used that do not need to be permanently marked in the forest. However, a marking that remains for a few months is recommended in case some sampling units need to be remediated.

They can also be sporadic. There are forest owners who carry out inventories only when they want to update stock volumes for accounting reasons, for example, or when they intend to use the wood.

7.1.4.2 Continuous inventories

Continuous inventories, also known as forest monitoring, are carried out on multiple correlated occasions and are used to analyze growth, forest evolution and medium and long-term planning.

They are carried out on permanent sampling units which are marked out in the field with highly durable materials such as oil paint, permanently demarcating their boundaries and numbering all the trees with the same number throughout the life of the tree.

7.2 Pilot Inventory

The pilot inventory is a small sample with the following objectives:

- Test the chosen methodology;
- Determine the time required for movement, demarcation and measurement;
- Define the equipment and materials needed;
- Training the inventory team;
- Determine the preliminary mean and variance;
- Determine the number of sample units needed to represent the population with a given probability of confidence for the mean.

The process to be used in the pilot inventory must have the same methodology as the definitive inventory.

7.3 Inventory plan

Inventory planning comes next, after the pilot inventory and determining the number of sample plots needed for the final inventory with the desired accuracy.

The planning should result in a document that includes the following topics:

- Introduction;
- Objectives;
- Literature review;

- Methodology;
- Activities;
  - Field survey;
  - Data processing;
  - Analysis of results;
  - Report preparation;
  - Presentation/dissemination of results;
- Timeline;
- Budget.

7.4 Team training

Training for the inventory team should be theoretical and practical and carried out during the pilot inventory on the distribution of AUs, calibration and operation of equipment, location, demarcation and measurement of sampling units, individuals to consider and variables to measure.

7.5 Carrying out the survey

Carrying out the forest inventory involves the following activities:

- Mapping;
- Distribution of UAs;
- Work plan - List of tasks, personnel, equipment;
- Equipment calibration;
- Locating, demarcating and measuring the AUs;
- Measurement of individuals in the UAs;
- Downloading data;

- Checking and correcting data.

7.5.1 Phases of timber forest inventory execution

A timber forest inventory is carried out in the following sequence of activities:

- Mapping;
- Distribution of AUs and trees to be felled;
- Work plan - List of tasks, personnel, equipment;
- UAs;
  - Equipment calibration;
  - Locating, demarcating and measuring the AUs;
  - Measurement of individuals in the UAs;
  - Downloading data;
  - Checking and correcting data;
- Cubage;
  - Equipment calibration;
  - Preparing tools (chainsaw, machete);
  - Locating the sampling points for tree selection;
  - Felling and measuring trees;
  - Data entry;
  - Checking and correcting data;
  - Calculating volumes;
  - Adjustment of equations.

7.5.2 Tools, equipment and materials used

The main equipment and tools used in forest inventories are listed below:

- Chainsaw;

- Sharpening guide;
- Boring lime;
- Limatão (belt);
- Adjustment key;
- Lubricating oil;
- Fuel;
- Machete;
- Scythe;
- Lever;
- Wedges;
- PPE - gloves, visor, helmet, ear protection, steel-toed boots, operator pants (12 ply);
- Location map of the AUs and trees for cubing;
- Clipboard, pen, pencil, eraser and survey sheets;
- Data collector or tablet;
- GPS;
- Hypsometer;
- Set of batteries for the devices and spare set;
- Compass;
- Trena/Suta for CAP/DAP;
- 50 m tape measure;
- Beacons;
- Crosshead;
- Canopy radius meter;
- Transport vehicle.

7.5.3 Inventory team

A forest inventory team should consist of:

- Coordinator;

- Diameter gauge;
- Height gauge;
- Note-taker;
- General assistant.

7.6 Processing data

Data processing consists of the following:

- Classification and structuring;
- adjustment of equations:
  - distribution/frequency;
  - growth;
  - sites / basal area;
  - hypsometric relationship;
  - volume;
  - assortments;
- Program;
- Processing itself.

7.7 Analyzing results

Analyzing the results of an inventory involves:

- Compare with previous and similar ones ;
- Look for causes of differences;
- Making predictions;
- Determine statistics for planning.

7.8 Report

The forest inventory report is the subject of Chapter 8 and should consist of the following:

- Executive summary;
- Summary;
- Introduction;
  - Identification;
  - Background;
  - Literature review;
    - oAbout the object of the inventory;
    - oOn other similar inventories;
    - oOn the results of inventories in the region.. ;
  - Description of the current situation;
- Objectives;
- Methodology;
  - Object (Forest and Environment);
  - Accuracy required;
  - Time approach;
  - Sampling process;
  - Sample units;
  - Measured variables;
  - Estimated variables;
  - Measurements taken and equipment used;
  - Statistics performed;
  - Equations tested;
  - Others..;
- Results and discussion;
- Conclusions;
- Bibliography cited

7.9 Presenting and disseminating results

The presentation and dissemination of the forest inventory results goes through the following stages:

- Printing the report (1st version);
- Distribution/delivery of the report;
- Preparation of the presentation;
- Presentation;
- Corrections;
- Printing the final version;
- Distribution of the final report.

8. FOREST INVENTORY REPORT

The report on a forest inventory is generally technical in nature, but can also include managerial aspects relating to logistics, execution time, etc. Here we will stick to the technical aspect as it is more comprehensible to students of the subject and makes it easier to teach.

8.1 Components of a forest inventory report

An example of a forest inventory report is shown in Appendix A.

The parts of a forest inventory technical report should include:

1) Cover;
2) Title page;
3) Presentation - explain the reason for the inventory;
4) Executive summary - a useful summary for those who only need an overview of the inventory;
5) Table of contents - level 1 and 2 headings with page references;
6) Introduction - history of the area to be inventoried, social and economic environment, description of the structure of the report, etc;
7) Objectives - identify the general and specific objectives of the inventory;
8) Methodology - identification of the forest, site and environment; temporal approach to the inventory; sampling method, system and process, including

equations used for the different stages and purposes of the inventory, required precision, etc.; team carrying out the pilot and definitive inventory; materials and equipment used; data processing method;

9) Pilot inventory - time to access the forest; time to locate the sampling units; average time needed to measure each sampling unit; statistical results and determination of sampling sufficiency;
10) Definitive inventory - inventory results in terms of the number of sampling units installed and measured; trunk cubing and analysis; equation adjustment and modeling; site classification; distribution by diametric class; sampling statistics; discussion of results;
11) Conclusions on the general and specific objectives;
12) Recommendations;
13) References;
14) Annexes and appendices.

8.2 Example of a forest inventory report

A complete example of a forest inventory report is presented in Appendix 1.

8.2.1 Cover

Both the cover and the title page, or the structure of a forest inventory report may vary according to the standards used by each professional or executing organization, or the requirements of the contractor. Therefore, consider this example report as a suggestion.

The cover should contain the name of the organization or professional who carried out the inventory at the top, in the middle should be the title of the report and at the foot of the page the location overlaying the year of execution (Figura 53 - a).

8.2.2 Title page

The title of the report must appear in the header of the cover sheet, then the names of the representatives of the contractor or client of the service. Below that, the name of the executing organization, if any, and the list of technical personnel on the executing team. At the foot of the page, again the location overlapping the year of execution (Figura 53 - b).

8.2.3 Presentation sheet

There won't always be a presentation; if there is, it should be to explain why the inventory is necessary in a few paragraphs and on a single page (Figura 53 - c).

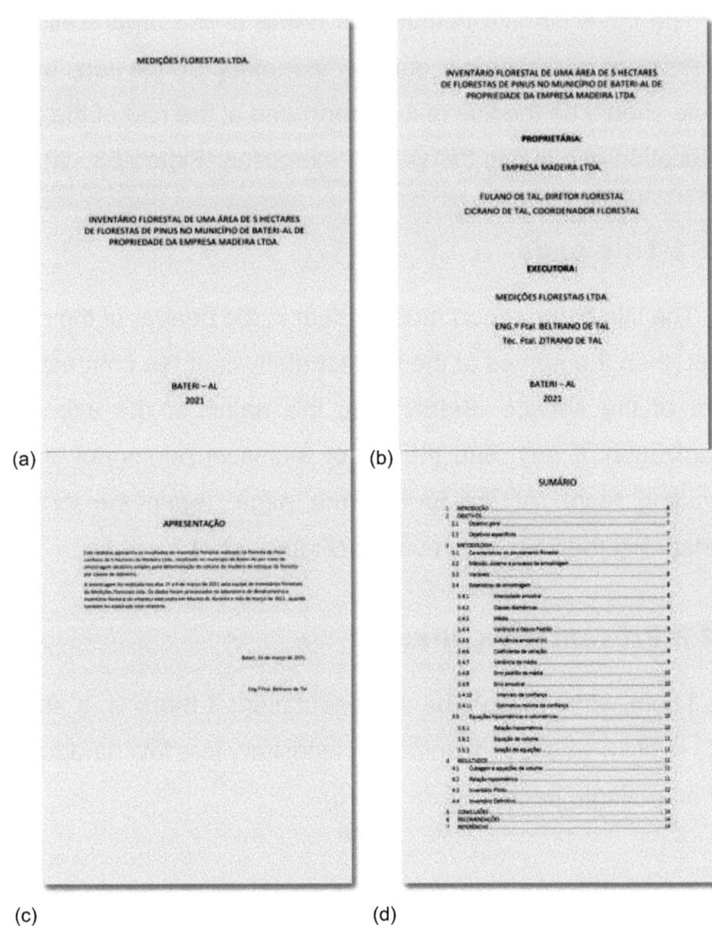

MEDIÇÕES FLORESTAIS LTDA

INVENTÁRIO FLORESTAL DE UMA ÁREA DE 5 HECTARES DE FLORESTAS DE PINUS NO MUNICÍPIO DE BATERI-AL DE PROPRIEDADE DA EMPRESA MADEIRA LTDA

BATERI – AL
2021

(a)

INVENTÁRIO FLORESTAL DE UMA ÁREA DE 5 HECTARES DE FLORESTAS DE PINUS NO MUNICÍPIO DE BATERI-AL DE PROPRIEDADE DA EMPRESA MADEIRA LTDA

PROPRIETÁRIA:

EMPRESA MADEIRA LTDA

FULANO DE TAL, DIRETOR FLORESTAL
CICRANO DE TAL, COORDENADOR FLORESTAL

EXECUTORA:

MEDIÇÕES FLORESTAIS LTDA

ENG.º Ftal. BELTRANO DE TAL
Téc. Ftal. ZITRANO DE TAL

BATERI – AL
2021

(b)

APRESENTAÇÃO

(c)

SUMÁRIO

(d)

FIGURA 53 - Example of the cover page (a), title page (b), presentation page (c) and summary (d) of a forest inventory report.

8.2.4 Executive summary

The executive summary should not exceed one page and should briefly describe the area inventoried, the purpose of the inventory, the methodology used, the results and conclusions, with the aim of being useful for those who only need a summary of the inventory. From this point on, the report refers to the technical part itself.

8.2.5 Summary

This is the list of level 1 and 2 titles with the references of the pages where they are found (Figura 53 - d).

8.2.6 Body of the report

The body of the report consists of

1. Introduction
2. Objectives
3. Methodology
4. Results and discussion
5. Conclusions
6. Recommendations
7. References

8.2.7 Annexes and appendices

Annexes are third-party documents that the forest inventory executors deem necessary for a perfect understanding and complementation of the information contained in the report.

Appendices are documents such as data tables, regression analysis results, etc., which do not need to be included in the body of the report, but are an integral part of the work carried out.

9. REFERENCES

COCHRAN, W. G. **Sampling Techniques**. Rio de Janeiro: USAID / Fundo de Cultura, 1965. 555p. Translation by Fernando A. Moreira Barbosa: "*COCHRAN, W. G. Sampling techniques, 2ed. John Wiley & Sons, 1953.*"

DRUSZCZ, J. P. et al. Efficiency of forest inventory with Bitterlich point sampling and line conglomerate in a *Pinus taeda* plantation. **FLORESTA**, Curitiba, PR, v. 42, n. 3, p. 527 - 538, jul./set. 2012.

FLORIANO, E. P. **Forest Phytosociology**. São Gabriel: ed. by the author, 2014.

FLORIANO, E. P. **Forest management**: for sustainability and excellence. Rio Largo: ed. by the author, 2018.

PÉLLICO NETTO, S.; BRENA, D. A. **Inventário Florestal**. Curitiba, 1997. 316 p.

RETSLAFF, F. A. S. et al. Conglomerate sampling by the Bitterlich method in Mixed Ombrophilous Forest. **Nativa**, Sinop, v. 02, n. 04, p. 194-198, Oct./Dec. 2014.

SAS. SAS/STAT ® 14.2 User's Guide. Cary, NC: SAS Institute Inc, 2016.

10. Appendix A - Cubing data for 24 trees

Tree	Section	d	e	h	di	ei	2ei	compi	s	hi	vi	dsi	vsi
1	0	29,0	1,0	27,4	38,5	1,7	3,4	0,11	t	0,11	0,012806	35,1	0,010644
1	1	29,0	1,0	27,4	27,0	0,8	1,6	2,50	s3	2,61	0,194828	25,4	0,384799
1	2	29,0	1,0	27,4	26,0	0,5	1,0	2,50	s3	5,11	0,137886	25,0	0,509764
1	3	29,0	1,0	27,4	26,0	0,4	0,8	2,50	s3	7,61	0,132732	25,2	0,753111
1	4	29,0	1,0	27,4	24,5	0,5	1,0	2,50	s2	10,11	0,125185	23,5	0,942753
1	5	29,0	1,0	27,4	22,5	0,4	0,8	2,50	s2	12,61	0,108434	21,7	1,013305
1	6	29,0	1,0	27,4	20,0	0,3	0,6	2,50	s2	15,11	0,088664	19,4	1,005463
1	7	29,0	1,0	27,4	19,5	0,3	0,6	2,10	s1	17,21	0,064334	18,9	0,991545
1	8	29,0	1,0	27,4	16,0	0,2	0,4	2,10	s1	19,31	0,051964	15,6	0,910827
1	9	29,0	1,0	27,4	13,0	0,2	0,4	2,10	m	21,41	0,034677	12,6	0,676180
2	0	32,5	2,2	25,4	38,0	2,6	5,2	0,11	t	0,11	0,012475	32,8	0,009295
2	1	32,5	2,2	25,4	30,0	1,2	2,4	2,50	s4	2,61	0,220353	27,6	0,376688
2	2	32,5	2,2	25,4	27,0	0,9	1,8	2,50	s3	5,11	0,159485	25,2	0,560590
2	3	32,5	2,2	25,4	26,0	0,7	1,4	2,50	s3	7,61	0,137886	24,6	0,741253
2	4	32,5	2,2	25,4	23,5	0,4	0,8	2,50	s2	10,11	0,120276	22,7	0,889679
2	5	32,5	2,2	25,4	21,0	0,2	0,4	2,50	s2	12,61	0,097205	20,6	0,930617
2	6	32,5	2,2	25,4	19,0	0,2	0,4	2,10	s1	14,71	0,065973	18,6	0,889967
2	7	32,5	2,2	25,4	16,5	0,1	0,2	2,10	s1	16,81	0,051964	16,3	0,807534
2	8	32,5	2,2	25,4	12,5	0,2	0,4	2,10	m	18,91	0,034677	12,1	0,612046
2	9	32,5	2,2	25,4	8,5	0,1	0,2	2,10	m	21,01	0,018184	8,3	0,355271
3	0	38,0	2,6	27,8	46,0	4,0	8,0	0,20	t	0,20	0,033238	38,0	0,022682
3	1	38,0	2,6	27,8	37,0	2,0	4,0	2,50	s4	2,70	0,319417	33,0	0,537142
3	2	38,0	2,6	27,8	34,0	1,2	2,4	2,50	s4	5,20	0,247450	31,6	0,852574
3	3	38,0	2,6	27,8	32,0	1,0	2,0	2,50	s4	7,70	0,213825	30,0	1,148167
3	4	38,0	2,6	27,8	29,5	0,7	1,4	2,50	s3	10,20	0,185661	28,1	1,353557
3	5	38,0	2,6	27,8	28,0	0,6	1,2	2,50	s3	12,70	0,162295	26,8	1,504014
3	6	38,0	2,6	27,8	25,0	0,5	1,0	2,50	s3	15,20	0,137886	24,0	1,545070
3	7	38,0	2,6	27,8	22,5	0,4	0,8	2,50	s2	17,70	0,110753	21,7	1,455339
3	8	38,0	2,6	27,8	18,0	0,3	0,6	2,10	s1	19,80	0,067633	17,4	1,203094
3	9	38,0	2,6	27,8	12,0	0,1	0,2	2,10	m	21,90	0,037110	11,8	0,760250
4	0	22,5	0,9	24,3	27,0	2,2	4,4	0,22	t	0,22	0,012596	22,6	0,008825
4	1	22,5	0,9	24,3	22,5	0,8	1,6	2,50	s3	2,72	0,113097	20,9	0,202428
4	2	22,5	0,9	24,3	21,5	0,7	1,4	2,50	s2	5,22	0,095033	20,1	0,344718
4	3	22,5	0,9	24,3	19,0	0,6	1,2	2,10	s1	7,32	0,067633	17,8	0,414425
4	4	22,5	0,9	24,3	18,0	0,4	0,8	2,10	s1	9,42	0,056449	17,2	0,453288
4	5	22,5	0,9	24,3	17,5	0,3	0,6	2,10	s1	11,52	0,051964	16,9	0,526084
4	6	22,5	0,9	24,3	16,0	0,3	0,6	2,10	s1	13,62	0,046274	15,4	0,559213
4	7	22,5	0,9	24,3	14,0	0,4	0,8	2,10	m	15,72	0,037110	13,2	0,507933
4	8	22,5	0,9	24,3	13,0	0,2	0,4	2,10	m	17,82	0,030059	12,6	0,466060
4	9	22,5	0,9	24,3	8,5	0,1	0,2	2,10	m	19,92	0,019060	8,3	0,356161

5	0	37,0	1,4	28,6	49,5	2,0	4,0	0,14	t	0,14	0,026942	45,5	0,022764
5	1	37,0	1,4	28,6	35,5	1,0	2,0	2,50	s4	2,64	0,324719	33,5	0,661949
5	2	37,0	1,4	28,6	34,5	0,9	1,8	2,50	s4	5,14	0,240528	32,7	0,884713
5	3	37,0	1,4	28,6	33,0	0,7	1,4	2,50	s4	7,64	0,223654	31,6	1,240801
5	4	37,0	1,4	28,6	31,0	0,5	1,0	2,50	s4	10,14	0,201062	30,0	1,512001
5	5	37,0	1,4	28,6	29,0	1,0	2,0	2,50	s3	12,64	0,176715	27,0	1,617179
5	6	37,0	1,4	28,6	28,0	0,5	1,0	2,50	s3	15,14	0,159485	27,0	1,733697
5	7	37,0	1,4	28,6	25,0	0,3	0,6	2,50	s3	17,64	0,137886	24,4	1,834824
5	8	37,0	1,4	28,6	22,0	0,1	0,2	2,50	s2	20,14	0,108434	21,8	1,693466
5	9	37,0	1,4	28,6	18,0	0,1	0,2	2,10	s1	22,24	0,065973	17,8	1,383546
5	10	37,0	1,4	28,6	11,0	0,1	0,2	2,10	m	24,34	0,034677	10,8	0,828666
6	0	18,5	1,2	18,6	30,0	2,0	4,0	0,06	t	0,06	0,004241	26,0	0,003186
6	1	18,5	1,2	18,6	17,5	0,8	1,6	2,10	s1	2,16	0,079828	15,9	0,157569
6	2	18,5	1,2	18,6	17,0	0,6	1,2	2,10	s1	4,26	0,049078	15,8	0,168110
6	3	18,5	1,2	18,6	15,0	0,4	0,8	2,10	s1	6,36	0,042223	14,2	0,225420
6	4	18,5	1,2	18,6	14,5	0,4	0,8	2,10	m	8,46	0,035883	13,7	0,258689
6	5	18,5	1,2	18,6	13,5	0,2	0,4	2,10	m	10,56	0,032327	13,1	0,297996
6	6	18,5	1,2	18,6	11,0	0,1	0,2	2,10	m	12,66	0,024750	10,8	0,286611
7	0	34,0	2,0	26,2	44,0	3,0	6,0	0,13	t	0,13	0,019767	38,0	0,014743
7	1	34,0	2,0	26,2	32,5	1,8	3,6	2,50	s4	2,63	0,266386	28,9	0,470793
7	2	34,0	2,0	26,2	30,5	1,0	2,0	2,50	s4	5,13	0,194828	28,5	0,663777
7	3	34,0	2,0	26,2	29,0	0,8	1,6	2,50	s3	7,63	0,173782	27,4	0,936647
7	4	34,0	2,0	26,2	28,5	0,8	1,6	2,50	s3	10,13	0,162295	26,9	1,173021
7	5	34,0	2,0	26,2	25,0	0,7	1,4	2,50	s3	12,63	0,140500	23,6	1,270272
7	6	34,0	2,0	26,2	23,0	0,6	1,2	2,50	s2	15,13	0,113097	21,8	1,226571
7	7	34,0	2,0	26,2	19,0	0,3	0,6	2,10	s1	17,23	0,072736	18,4	1,101268
7	8	34,0	2,0	26,2	15,0	0,2	0,4	2,10	m	19,33	0,047666	14,6	0,837607
7	9	34,0	2,0	26,2	10,0	0,1	0,2	2,10	m	21,43	0,025771	9,8	0,520417
8	0	39,0	2,5	26,8	51,0	3,5	7,0	0,15	t	0,15	0,030642	44,0	0,022808
8	1	39,0	2,5	26,8	35,5	1,0	2,0	2,50	s4	2,65	0,343617	33,5	0,636515
8	2	39,0	2,5	26,8	35,0	1,2	2,4	2,50	s4	5,15	0,243977	32,6	0,883793
8	3	39,0	2,5	26,8	33,5	0,9	1,8	2,50	s4	7,65	0,230330	31,7	1,242305
8	4	39,0	2,5	26,8	32,0	0,9	1,8	2,50	s4	10,15	0,210597	30,2	1,528137
8	5	39,0	2,5	26,8	30,5	0,6	1,2	2,50	s4	12,65	0,191748	29,3	1,759072
8	6	39,0	2,5	26,8	29,0	0,5	1,0	2,50	s3	15,15	0,173782	28,0	1,954363
8	7	39,0	2,5	26,8	24,5	0,4	0,8	2,50	s2	17,65	0,140500	23,7	1,865433
8	8	39,0	2,5	26,8	17,5	0,3	0,6	2,10	s1	19,75	0,072736	16,9	1,314299
8	9	39,0	2,5	26,8	12,0	0,1	0,2	2,10	m	21,85	0,035883	11,8	0,729083
9	0	38,5	2,0	27,0	50,0	3,0	6,0	0,10	t	0,10	0,019635	44,0	0,015205
9	1	38,5	2,0	27,0	35,5	1,5	3,0	2,50	s4	2,60	0,335452	32,5	0,611028
9	2	38,5	2,0	27,0	33,0	1,0	2,0	2,50	s4	5,10	0,230330	31,0	0,808016
9	3	38,5	2,0	27,0	33,0	0,9	1,8	2,50	s4	7,60	0,213825	31,2	1,154672
9	4	38,5	2,0	27,0	32,0	0,8	1,6	2,50	s4	10,10	0,207394	30,4	1,505275
9	5	38,5	2,0	27,0	29,0	0,5	1,0	2,50	s3	12,60	0,182654	28,0	1,690398
9	6	38,5	2,0	27,0	26,0	0,4	0,8	2,50	s3	15,10	0,148489	25,2	1,682912
9	7	38,5	2,0	27,0	23,0	0,3	0,6	2,50	s2	17,60	0,117859	22,4	1,571400

9	8	38,5	2,0	27,0	18,0	0,2	0,4	2,10	s1	19,70	0,069313	17,6	1,255612
9	9	38,5	2,0	27,0	12,0	0,1	0,2	2,10	m	21,80	0,037110	11,8	0,768763
10	0	33,0	1,8	22,6	43,0	2,0	4,0	0,08	t	0,08	0,011618	39,0	0,009557
10	1	33,0	1,8	22,6	29,0	1,5	3,0	2,50	s3	2,58	0,240528	26,0	0,445184
10	2	33,0	1,8	22,6	29,0	1,2	2,4	2,50	s3	5,08	0,165130	26,6	0,552016
10	3	33,0	1,8	22,6	28,0	1,0	2,0	2,50	s3	7,58	0,159485	26,0	0,823677
10	4	33,0	1,8	22,6	25,0	0,8	1,6	2,50	s3	10,08	0,137886	23,4	0,968670
10	5	33,0	1,8	22,6	23,5	0,7	1,4	2,50	s2	12,58	0,115466	22,1	1,023570
10	6	33,0	1,8	22,6	20,0	0,5	1,0	2,50	s2	15,08	0,092886	19,0	1,006025
10	7	33,0	1,8	22,6	13,0	0,3	0,6	2,10	m	17,18	0,044903	12,4	0,694573
10	8	33,0	1,8	22,6	8,0	0,1	0,2	2,10	m	19,28	0,018184	7,8	0,324958
11	0	33,5	2,0	20,2	47,0	3,0	6,0	0,10	t	0,10	0,017349	41,0	0,013203
11	1	33,5	2,0	20,2	33,0	1,5	3,0	2,50	s4	2,60	0,281047	30,0	0,527049
11	2	33,5	2,0	20,2	31,5	1,0	2,0	2,50	s4	5,10	0,204216	29,5	0,709079
11	3	33,5	2,0	20,2	28,5	0,9	1,8	2,50	s3	7,60	0,176715	26,7	0,944980
11	4	33,5	2,0	20,2	27,0	0,7	1,4	2,50	s3	10,10	0,151201	25,6	1,085367
11	5	33,5	2,0	20,2	24,0	0,5	1,0	2,50	s2	12,60	0,127676	23,0	1,172045
11	6	33,5	2,0	20,2	14,0	0,3	0,6	2,10	m	14,70	0,059541	13,4	0,818058
11	7	33,5	2,0	20,2	8,0	0,1	0,2	2,10	m	16,80	0,019957	7,8	0,317200
12	0	25,5	1,4	20,2	32,5	2,5	5,0	0,13	t	0,13	0,010784	27,5	0,007721
12	1	25,5	1,4	20,2	23,0	1,0	2,0	2,50	s3	2,63	0,143139	21,0	0,247304
12	2	25,5	1,4	20,2	23,0	0,7	1,4	2,50	s3	5,13	0,103869	21,6	0,365664
12	3	25,5	1,4	20,2	21,0	0,4	0,8	2,50	s3	7,63	0,095033	20,2	0,524112
12	4	25,5	1,4	20,2	20,0	0,3	0,6	2,50	s3	10,13	0,082516	19,4	0,624075
12	5	25,5	1,4	20,2	18,5	0,3	0,6	2,10	s1	12,23	0,061118	17,9	0,669277
12	6	25,5	1,4	20,2	15,5	0,2	0,4	2,10	s1	14,33	0,047666	15,1	0,617233
12	7	25,5	1,4	20,2	9,5	0,1	0,2	2,10	m	16,43	0,025771	9,3	0,405834
13	0	28,0	1,0	24,0	38,5	2,0	4,0	0,08	t	0,08	0,009313	34,5	0,007479
13	1	28,0	1,0	24,0	27,5	0,7	1,4	2,50	s3	2,58	0,192772	26,1	0,379219
13	2	28,0	1,0	24,0	26,5	0,6	1,2	2,50	s3	5,08	0,143139	25,3	0,527175
13	3	28,0	1,0	24,0	25,0	0,5	1,0	2,50	s3	7,58	0,130192	24,0	0,723977
13	4	28,0	1,0	24,0	23,5	0,4	0,8	2,50	s2	10,08	0,115466	22,7	0,863954
13	5	28,0	1,0	24,0	21,0	0,2	0,4	2,50	s2	12,58	0,097205	20,6	0,928403
13	6	28,0	1,0	24,0	19,0	0,2	0,4	2,10	s1	14,68	0,065973	18,6	0,888152
13	7	28,0	1,0	24,0	14,5	0,1	0,2	2,10	m	16,78	0,046274	14,3	0,725437
14	0	28,0	1,3	22,1	36,0	3,0	6,0	0,09	t	0,09	0,009161	30,0	0,006362
14	1	28,0	1,3	22,1	29,0	1,0	2,0	2,50	s3	2,59	0,188692	27,0	0,331368
14	2	28,0	1,3	22,1	25,5	1,0	2,0	2,50	s3	5,09	0,145802	23,5	0,512202
14	3	28,0	1,3	22,1	25,0	0,9	1,8	2,50	s3	7,59	0,125185	23,2	0,650060
14	4	28,0	1,3	22,1	23,5	0,6	1,2	2,50	s2	10,09	0,115466	22,3	0,820623
14	5	28,0	1,3	22,1	20,5	0,4	0,8	2,50	s2	12,59	0,095033	19,7	0,875478
14	6	28,0	1,3	22,1	17,5	0,3	0,6	2,10	s1	14,69	0,059541	16,9	0,777281
14	7	28,0	1,3	22,1	13,0	0,1	0,2	2,10	m	16,79	0,038357	12,8	0,592682
15	0	23,0	1,3	20,7	31,0	2,5	5,0	0,10	t	0,10	0,007548	26,0	0,005309
15	1	23,0	1,3	20,7	21,0	1,0	2,0	2,50	s2	2,60	0,122718	19,0	0,211759
15	2	23,0	1,3	20,7	20,0	0,6	1,2	2,50	s2	5,10	0,082516	18,8	0,286171

15	3	23,0	1,3	20,7	20,0	0,5	1,0	2,50	s2	7,60	0,078540	19,0	0,426451
15	4	23,0	1,3	20,7	18,0	0,4	0,8	2,10	s1	9,70	0,059541	17,2	0,500405
15	5	23,0	1,3	20,7	15,5	0,3	0,6	2,10	s1	11,80	0,046274	14,9	0,479928
15	6	23,0	1,3	20,7	13,0	0,3	0,6	2,10	m	13,90	0,033492	12,4	0,410229
15	7	23,0	1,3	20,7	9,0	0,1	0,2	2,10	m	16,00	0,019957	8,8	0,290534
16	0	22,5	2,0	22,7	28,0	1,5	3,0	0,08	t	0,08	0,004926	25,0	0,003927
16	1	22,5	2,0	22,7	21,5	1,0	2,0	2,50	s2	2,58	0,113097	19,5	0,203697
16	2	22,5	2,0	22,7	20,0	0,7	1,4	2,50	s2	5,08	0,084541	18,6	0,289745
16	3	22,5	2,0	22,7	19,0	0,4	0,8	2,10	s1	7,18	0,062716	18,2	0,381884
16	4	22,5	2,0	22,7	18,0	0,3	0,6	2,10	s1	9,28	0,056449	17,4	0,462091
16	5	22,5	2,0	22,7	16,0	0,2	0,4	2,10	s1	11,38	0,047666	15,6	0,488113
16	6	22,5	2,0	22,7	13,5	0,1	0,2	2,10	m	13,48	0,035883	13,3	0,444926
16	7	22,5	2,0	22,7	10,0	0,1	0,2	2,10	m	15,58	0,022771	9,8	0,333971
17	0	21,5	1,0	22,0	32,5	2,5	5,0	0,08	t	0,08	0,006637	27,5	0,004752
17	1	21,5	1,0	22,0	20,5	0,4	0,8	2,50	s2	2,58	0,121088	19,7	0,231881
17	2	21,5	1,0	22,0	20,5	0,8	1,6	2,50	s2	5,08	0,082516	18,9	0,297361
17	3	21,5	1,0	22,0	20,0	0,7	1,4	2,50	s2	7,58	0,080516	18,6	0,418619
17	4	21,5	1,0	22,0	17,5	0,4	0,8	2,10	s1	9,68	0,057984	16,7	0,475052
17	5	21,5	1,0	22,0	16,5	0,5	1,0	2,10	s1	11,78	0,047666	15,5	0,480308
17	6	21,5	1,0	22,0	14,0	0,3	0,6	2,10	m	13,88	0,038357	13,4	0,457649
17	7	21,5	1,0	22,0	10,5	0,2	0,4	2,10	m	15,98	0,024750	10,1	0,353389
18	0	16,0	0,9	18,0	22,5	1,5	3,0	0,08	t	0,08	0,003181	19,5	0,002389
18	1	16,0	0,9	18,0	15,0	0,7	1,4	2,10	s1	2,18	0,052453	13,6	0,096773
18	2	16,0	0,9	18,0	13,0	0,5	1,0	2,10	m	4,28	0,032327	12,0	0,110580
18	3	16,0	0,9	18,0	12,0	0,2	0,4	2,10	m	6,38	0,025771	11,6	0,139582
18	4	16,0	0,9	18,0	9,5	0,2	0,4	2,10	m	8,48	0,019060	9,1	0,144772
18	5	16,0	0,9	18,0	8,0	0,1	0,2	2,10	m	10,58	0,012628	7,8	0,119366
19	0	17,0	0,9	19,0	21,0	2,0	4,0	0,08	t	0,08	0,002771	17,0	0,001816
19	1	17,0	0,9	19,0	16,0	0,8	1,6	2,10	s1	2,18	0,053438	14,4	0,084985
19	2	17,0	0,9	19,0	15,0	0,4	0,8	2,10	s1	4,28	0,039625	14,2	0,137486
19	3	17,0	0,9	19,0	15,0	0,3	0,6	2,10	s1	6,38	0,037110	14,4	0,204943
19	4	17,0	0,9	19,0	12,0	0,2	0,4	2,10	m	8,48	0,030059	11,6	0,227725
19	5	17,0	0,9	19,0	10,5	0,1	0,2	2,10	m	10,58	0,020874	10,3	0,199968
19	6	17,0	0,9	19,0	9,0	0,1	0,2	2,10	m	12,68	0,015679	8,8	0,182775
20	0	17,0	1,0	21,3	24,0	1,5	3,0	0,09	t	0,09	0,004072	21,0	0,003117
20	1	17,0	1,0	21,3	16,5	0,9	1,8	2,10	s1	2,19	0,060590	14,7	0,113021
20	2	17,0	1,0	21,3	15,5	0,5	1,0	2,10	s1	4,29	0,042223	14,5	0,143649
20	3	17,0	1,0	21,3	15,0	0,4	0,8	2,10	s1	6,39	0,038357	14,2	0,206715
20	4	17,0	1,0	21,3	14,0	0,3	0,6	2,10	m	8,49	0,034677	13,4	0,254185
20	5	17,0	1,0	21,3	12,5	0,2	0,4	2,10	m	10,59	0,028956	12,1	0,271121
20	6	17,0	1,0	21,3	12,0	0,1	0,2	2,10	m	12,69	0,024750	11,8	0,284699
20	7	17,0	1,0	21,3	10,0	0,2	0,4	2,10	m	14,79	0,019957	9,6	0,268795
21	0	13,0	1,0	13,0	18,0	2,0	4,0	0,09	t	0,09	0,002290	14,0	0,001385
21	1	13,0	1,0	13,0	12,0	0,9	1,8	2,10	m	2,19	0,033885	10,2	0,051608
21	2	13,0	1,0	13,0	11,0	0,5	1,0	2,10	m	4,29	0,021812	10,0	0,068748
21	3	13,0	1,0	13,0	9,0	0,3	0,6	2,10	m	6,39	0,016493	8,4	0,085599

22	0	14,0	0,7	18,6	23,0	2,0	4,0	0,07	t	0,07	0,002908	19,0	0,001985
22	1	14,0	0,7	18,6	13,5	0,3	0,6	2,10	m	2,17	0,046736	12,9	0,089887
22	2	14,0	0,7	18,6	12,0	0,2	0,4	2,10	m	4,27	0,026812	11,6	0,100935
22	3	14,0	0,7	18,6	11,5	0,1	0,2	2,10	m	6,37	0,022771	11,3	0,131203
22	4	14,0	0,7	18,6	10,0	0,1	0,2	2,10	m	8,47	0,019060	9,8	0,148832
22	5	14,0	0,7	18,6	9,0	0,1	0,2	2,10	m	10,57	0,014885	8,8	0,144017
23	0	14,0	0,9	16,5	18,0	1,0	2,0	0,06	t	0,06	0,001527	16,0	0,001206
23	1	14,0	0,9	16,5	15,0	0,6	1,2	2,10	s1	2,16	0,040482	13,8	0,075737
23	2	14,0	0,9	16,5	13,0	0,4	0,8	2,10	m	4,26	0,032327	12,2	0,113516
23	3	14,0	0,9	16,5	10,0	0,2	0,4	2,10	m	6,36	0,021812	9,6	0,120383
23	4	14,0	0,9	16,5	9,5	0,1	0,2	2,10	m	8,46	0,015679	9,3	0,118703
24	0	13,0	0,5	17,6	18,5	1,5	3,0	0,08	t	0,08	0,002150	15,5	0,001510
24	1	13,0	0,5	17,6	13,0	0,4	0,8	2,10	m	2,18	0,036290	12,2	0,066619
24	2	13,0	0,5	17,6	11,0	0,3	0,6	2,10	m	4,28	0,023750	10,4	0,086391
24	3	13,0	0,5	17,6	10,0	0,2	0,4	2,10	m	6,38	0,018184	9,6	0,100377
24	4	13,0	0,5	17,6	8,5	0,1	0,2	2,10	m	8,48	0,014112	8,3	0,107262

Where: Tree - felled tree number; Section = tree section number; d = tree diameter at 1.3 m from the ground; e = simple bark thickness at 1.3 m from the ground; h = total tree height; di = trunk diameter at position i; ei = simple bark thickness at position i; 2ei = bark thickness multiplied by 2; Compi = length of section i; S = type of log assortment according to its diameter at the thin end (t = stump; s1 = log (<20cm); s2 = thin log (≥20-<25cm); s3 = medium log (≥25-<30cm; s4 = thick log (≥30cm); hi = tree height at position i; Vi = volume with bark of section i calculated by the Smalian method; dsi = diameter without bark at position i = d = 2ei; Vsi = volume without bark of section i calculated by the Smalian method.

11. Appendix B - h_{100} of the plots

Age	Location	Tranche	h100
5	1	1	5,433
5	1	2	5,633
5	1	3	4,900
5	1	4	4,933
5	2	1	4,667
5	2	2	4,767
5	2	3	4,833
5	2	4	4,867
5	3	1	4,533
5	3	2	4,833
5	3	3	4,800
5	3	4	4,767
5	4	1	4,933
5	4	2	4,967
5	4	3	5,033
5	4	4	5,133
5	5	1	5,033
5	5	2	5,033
5	5	3	5,167
5	5	4	5,267
6	1	1	7,000
6	1	2	7,233
6	1	3	6,300
6	1	4	6,333
6	2	1	6,000
6	2	2	6,100
6	2	3	6,200
6	2	4	6,267
6	3	1	5,833
6	3	2	6,200
6	3	3	6,167
6	3	4	6,133
6	4	1	6,333
6	4	2	6,367
6	4	3	6,433
6	4	4	6,600
6	5	1	6,467
6	5	2	6,467
6	5	3	6,633
6	5	4	6,767
7	1	1	8,767
7	1	2	9,033
7	1	3	7,900

Age	Location	Tranche	h100
7	1	4	7,933
7	2	1	7,467
7	2	2	7,633
7	2	3	7,767
7	2	4	7,833
7	3	1	7,267
7	3	2	7,733
7	3	3	7,700
7	3	4	7,633
7	4	1	7,933
7	4	2	7,967
7	4	3	8,033
7	4	4	8,267
7	5	1	8,067
7	5	2	8,100
7	5	3	8,300
7	5	4	8,467
8	1	1	10,667
8	1	2	11,000
8	1	3	9,600
8	1	4	9,633
8	2	1	9,100
8	2	2	9,267
8	2	3	9,433
8	2	4	9,533
8	3	1	8,867
8	3	2	9,400
8	3	3	9,367
8	3	4	9,300
8	4	1	9,633
8	4	2	9,667
8	4	3	9,767
8	4	4	10,033
8	5	1	9,833
8	5	2	9,833
8	5	3	10,100
8	5	4	10,300
9	1	1	12,700
9	1	2	13,067
9	1	3	11,400
9	1	4	11,433
9	2	1	10,833
9	2	2	11,000
9	2	3	11,200
9	2	4	11,333

Age	Location	Tranche	h100
9	3	1	10,567
9	3	2	11,167
9	3	3	11,133
9	3	4	11,067
9	4	1	11,433
9	4	2	11,467
9	4	3	11,600
9	4	4	11,900
9	5	1	11,700
9	5	2	11,667
9	5	3	12,000
9	5	4	12,233
10	1	1	14,800
10	1	2	15,233
10	1	3	13,300
10	1	4	13,333
10	2	1	12,600
10	2	2	12,833
10	2	3	13,067
10	2	4	13,200
10	3	1	12,300
10	3	2	13,000
10	3	3	12,967
10	3	4	12,867
10	4	1	13,333
10	4	2	13,367
10	4	3	13,500
10	4	4	13,867
10	5	1	13,600
10	5	2	13,600
10	5	3	13,967
10	5	4	14,267
11	1	1	16,933
11	1	2	17,400
11	1	3	15,200
11	1	4	15,233
11	2	1	14,367
11	2	2	14,667
11	2	3	14,933
11	2	4	15,100
11	3	1	14,033
11	3	2	14,867
11	3	3	14,833
11	3	4	14,733
11	4	1	15,233

Age	Location	Tranche	h100
11	4	2	15,267
11	4	3	15,433
11	4	4	15,833
11	5	1	15,567
11	5	2	15,533
11	5	3	15,967
11	5	4	16,300
12	1	1	19,067
12	1	2	19,567
12	1	3	17,100
12	1	4	17,133
12	2	1	16,133
12	2	2	16,500
12	2	3	16,800
12	2	4	17,000
12	3	1	15,767
12	3	2	16,733
12	3	3	16,700
12	3	4	16,600
12	4	1	17,133
12	4	2	17,167
12	4	3	17,367
12	4	4	17,800
12	5	1	17,533
12	5	2	17,467
12	5	3	17,967
12	5	4	18,333

12. Appendix C - Example of a forest inventory report

MEDIÇÕES FLORESTAIS LTDA.

INVENTÁRIO FLORESTAL DE UMA ÁREA DE 5 HECTARES DE FLORESTAS DE PINUS NO MUNICÍPIO DE BATERI-AL DE PROPRIEDADE DA EMPRESA MADEIRA LTDA.

BATERI – AL

2021

INVENTÁRIO FLORESTAL DE UMA ÁREA DE 5 HECTARES DE FLORESTAS DE PINUS NO MUNICÍPIO DE BATERI-AL DE PROPRIEDADE DA EMPRESA MADEIRA LTDA.

PROPRIETÁRIA:

EMPRESA MADEIRA LTDA.

FULANO DE TAL, DIRETOR FLORESTAL
CICRANO DE TAL, COORDENADOR FLORESTAL

EXECUTORA:

MEDIÇÕES FLORESTAIS LTDA.

ENG.º Ftal. BELTRANO DE TAL
Téc. Ftal. ZITRANO DE TAL

BATERI – AL
2021

APRESENTAÇÃO

Este relatório apresenta os resultados do inventário florestal realizado na floresta de *Pinus caribaea* de 5 hectares da Madeira Ltda., localizado no município de Bateri-AL por meio de amostragem aleatória simples para determinação do volume de madeira de estoque da floresta por classes de diâmetro.

A amostragem foi realizada nos dias 1º a 4 de março de 2021 pela equipe de inventários florestais da Medições Florestais Ltda. Os dados foram processados no laboratório de dendrometria e inventário florestal da empresa executora em Maceió-AL durante o mês de março de 2021, quando também foi elaborado este relatório.

Bateri, 31 de março de 2021.

Eng.º Ftal. Beltrano de Tal

RESUMO EXECUTIVO

A floresta de *Pinus caribaea* objeto deste inventário pertencente à Madeira Ltda. Encontra-se com 26 anos de idade e possui 5 hectares de efetivo plantio, localizados no município de Bateri-AL. O povoamento foi submetido a uma amostragem aleatória simples por meio de 9 parcelas amostrais que representam 9,1% da população. A floresta apresentou um número médio de 728 árvores por hectare (N), diâmetro médio de 19,9 cm e altura média de 14,4 m. A área basal média foi de 24,7 m^2/ha e o volume médio foi de 229,4 m^3/ha, sendo que mais de 70% do volume de madeira está concentrado nas árvores com diâmetro acima do médio. O volume total estimado é de 1147 m^3 de madeira nos 5 ha da floresta. O Erro Amostral é de 8% para mais ou para menos da média de volume por hectare. A estimativa minima de confiança para o volume total é de 1055 m^3.

SUMÁRIO

INTRODUÇÃO

A empresa Madeira Ltda. possui um pequeno povoamento de *Pinus cariboea* no município de Bateri-AL, com 26 anos de idade, do qual vem extraindo indivíduos para prover madeira para sua serraria localizada nas proximidades da floresta.

A empresa não depende do povoamento de pinus para abastecer a serraria e usa a madeira proveniente do mesmo para fins especiais esporádicos. Até o presente, tem realizado desbastes, retirando o necessário para essa demanda sem planejar as atividades. Recentemente, decidiu melhorar o manejo da floresta e solicitou a execução de um inventário florestal com essa finalidade, tendo contratado a empresa Medições Florestais Ltda. para executá-lo.

Por ser uma área pequena, decidiu-se por uma amostragem aleatória simples, que deverá mostrar os estoque por classes de diâmetro e a variabilidade da população, permitindo que se obtenha dados para definir práticas de manejo florestal para o futuro do povoamento florestal.

6

OBJETIVOS

Objetivo geral

Determinar o volume de estoque de madeira de um povoamento florestal de *Pinus cariboea* com 5 hectares pertencente à empresa Madeira Ltda., localizado no município de Bateri, AL.

Objetivos específicos

- **Ajustar equação de relação hipsométrica com os dados do inventário piloto para estimar a altura das árvores do inventário definitivo;**
- **Cubar ao menos 20 árvores e ajustar equação volumétrica para estimar o volume das árvores;**
- **Construir uma tabela de distribuição por classes de diâmetro;**
- **Determinar as estatísticas da amostragem de média, variância, variância da média, erro padrão da média e intervalo de confiança para a média com limite de erro amostral máximo de 10% e 95% de confiança para a média.**

METODOLOGIA

Características do povoamento florestal

A área se localiza no município de Bateri-AL, com coordenadas centrais de 9,92838º Sul e 35,99722º Oeste. O clima regional é tropical úmido, sem estação seca, com precipitação média anual de 1350 mm, com umidade relativa do ar acima de 40% e média de 77%, temperaturas mínima de 22ºC, máxima de 34ºC e média de 26ºC. O solo é franco arenoso, de baixa fertilidade, com pH em torno de 5,5. A topografia é ondulada, com altitudes variando de 10 a 65 metros.

O povoamento floretal constitui-se de uma pequena área de 5,0 há plantados com *Pinus cariboea* var. *hondurensis* (Sénécl) Barr. e Golf. de propriedade da empresa Madeira Ltda.

Atualmente a floresta apresenta 26 anos de idade. Foi plantada com espaçamento de 3 metros entre linhas de plantio e 2 metros entre plantas nas linhas, mas o espaçamento atual não apresenta muita regularidade, devido aos desbastes.

Método, sistema e processo de amostragem

As unidades amostrais utilizadas foram parcelas de área fixa, circulares, com área de 506,71 m², correspondendo a um raio de 12,7 m.

As parcelas foram distribuidas de forma aleatória sem restrições sobre a área, sendo inicialmente amostradas 6 parcelas no inventário piloto e, após a determinação do número necerrário para o inventário definitivo, este número foi ampliado para um total de 9 unidades amostrais.

A amostragem foi realizada conforme o processo de amostragem aleatória simples.

7

Variáveis

No inventário piloto foram cubadas 24 árvores, sendo medidos a altura do toco, os diâmetros na altura do toco, a 0,6 m de altura, os diâmetros a 1,3 metros de altura e a cada metro a partir daí, até o topo. Os volumes das árvores cubadas foram calculados pela equação de Smalian e os das árvores das parcelas foram estimados pela melhor equação de volume ajustada com os dados da cubagem.

No inventário piloto foram medidos os diâmetros a 1,3 metros de altura de todas as árvores das parcelas e as alturas das 10 primeiras para ajuste de relação hipsométrica. As alturas das demais árvores foram estimadas pela melhor equação ajustada.

No inventário definitivo, foram medidos somente os diâmetros a 1,3 m de altura de todas as árvores e estimadas todas as alturas e os volumes.

As áreas basais a 1,3 m de altura das árvores foram calculadas pela equação:

$$g = \pi \,.\, d^2 / 4$$

Onde: g = área basal individual da árvore em m²; d = diâmetro do tronco a 1,3 m de altura em metros.

O número de árvores por hectare (N) das parcelas foi encontrado pela multiplicação de 10000 pelo número de árvores da parcela amostral, divididos pela área da parcela em m².

A área basal média por hectare (G) das parcelas foi calculada pela multiplicação da área basal média individual da parcela (g) pelo número de árvores por hectare da parcela (N).

O volume médio por hectare (V) das parcelas foi calculado pela multiplicação do volume médio individual da parcela (v) pelo número de árvores por hectare da parcela (N).

Estatísticas da amostragem

As estatísticas da amostragem foram realizadas conforme recomendado por Cochran (1965).

Intensidade amostral (IA)

Intensidade amostral (IA) é representada pela percentagem da área da floresta em estudo que foi amostrada, sendo calculada por:

$$IA = n \,.\, a / A$$

Onde: IA = intensidade de amostragem em percentagem; n = número de unidades amostrais; a = área da unidade amostral em m²; A = área da floresta a inventariar em m².

No caso do inventário piloto, com 6 parcelas amostrais de 506,71 m², ou 3.040,26 m² da área de 50000 m² da floresta (5 ha), resultou numa intensidade amostral (IA) de 6,1%, o que significa que a população é considerada finita.

Classes diamétricas

O número de classes diamétricas foi determinado pelo método de Floriano (2018), utilizando-se a equação:

$$k = n^{0.175} \,.\, (LS_d - LI_d)^{0.3}$$

Onde: k = número de classes diamétricas; n = número total de árvores amostradas em todas as parcelas do inventário; LS_d = limite superior dos diâmetros das árvores em cm; LI_d = limite inferior dos diâmetros das árvores em cm.

Média

As médias amostrais foram calculadas pela expressão:

$$\bar{x} = \frac{\sum_{i=1}^{n} \bar{x}_i}{n}$$

Onde: $\bar{x}$ = média amostral; $\bar{x}_i$ = média da unidade amostral i; n = número de unidades amostrais; i = número de ordem da unidade amostral.

Variância e Desvio Padrão

A variância amostral foi calculada por:

$$s_x^2 = \frac{\sum_{i=1}^{n} (\bar{x}_i - \bar{x})^2}{n - 1}$$

Onde: s_x^2 = variância amostral; $\bar{x}$ = média amostral; $\bar{x}_i$ = média da unidade amostral i; n = número de unidades amostrais; i = número de ordem da unidade amostral.

O Desvio Padrão (S_x) foi calculado pela raiz quadrada da variância, como segue:

$$S_x = \sqrt{S_x^2}$$

Suficiência amostral (n)

A suficiência amostral é definida como o número de unidades amostrais suficientes para representar a população com o erro máximo admitido e previamente determinado em relação à média da população. O cálculo foi realizado para população finita com o fator de correção, como segue:

$$n = \frac{N . t_{(p,GL)}^2 . s_x^2}{N . E^2 + t_{(p,GL)}^2 . s_x^2}$$

Onde: n = número de unidades amostrais suficientes para representar a população; s_x^2 = variância amostral; $\bar{x}$ = média amostral; E = erro amostral admitido (E% x Média); N = número potencial total de parcelas da população; t(p,GL) = valor do t de Student para n-1 graus de liberdade (GL) e nível de probabilidade de confiança para a média desejado; n = número de unidades amostrais.

Coeficiente de variação

O coeficiente de variação em percentagem da média foi calculado pela expressão:

$$CV = \frac{100 . s_x}{\bar{x}}$$

Onde: CV = coeficiente de variação; s_x = desvio padrão; $\bar{x}$ = média amostral.

Variância da média

A Variância da Média foi determinada por:

$$s_{\bar{x}}^2 = \left(\frac{s_x^2}{n - 1}\right) . \left(\frac{N - n}{N}\right)$$

Onde: $S_{\bar{x}}^2$ = variância da média; S_x^2 = variância; n = número de unidades amostrais; N = número potencial de unidades amostrais; n = número de unidades amostrais.

Erro padrão da média

O Erro Padrão da Média foi determinado por:

$$S_{\bar{x}} = \pm \frac{s_x}{\sqrt{n}} \sqrt{\left(\frac{N-n}{N}\right)}$$

Onde: $S_{\bar{x}}$ = erro padrão da média; s_x = desvio padrão; n = número de unidades amostrais; N = número potencial de unidades amostrais.

Erro amostral

O Erro da Amostragem é calculado para valores absolutos e relativos, como segue:

Erro amostral absoluto (Ea)

$$E_a = \pm\ t_{(p,GL)} . S_{\bar{x}}$$

Erro amostral relativo (Er)

$$E_r = \pm\ \frac{100 . t_{(p,GL)} . S_{\bar{x}}}{\bar{x}}$$

Onde: Ea, Er = erro amostral absoluto e relativo, respectivamente; $S_{\bar{x}}$ = erro padrão da média; GL = n – 1 (graus de liberdade); p = probabilidade de confiança desejada para a média; t(p,GL) = valor do t de Student para n-1 graus de liberdade (GL) e probabilidade desejada; n = número de parcelas amostrais.

Intervalo de confiança

$$IC\ [(\bar{x} -\ t_{(p,GL)} . S_{\bar{x}}) \leq \mu \leq (\bar{x} +\ t_{(p,GL)} . S_{\bar{x}})] = p$$

Onde: IC = intervalo de confiança; μ = média da população; p = probabilidade de confiança desejada para a média; GL = graus de liberdade = n - 1; t(p,GL) = valor do t de Student para n-1 graus de liberdade (GL) e probabilidade desejada; n = número de parcelas amostrais.

Estimativa mínima de confiança

$$EMC = \bar{x} -\ t_{(p,GL)} . S_{\bar{x}}$$

Onde: EMC = estimativa mínima de confiança; $\bar{x}$ = média amostral; p = Probabilidade desejada de confiança para a média; GL = graus de liberdade (nº de unidades amostrais – 1); t(p,GL) = valor do t de Student para n-1 graus de liberdade (GL) e probabilidade desejada; n = número de parcelas amostrais.

Equações hipsométricas e volumétricas

Relação hipsométrica

Foram ajustados os seguintes modelos para relação hipsométrica:

- **h = b0 + b1 ln(d)**
- **h = b0 + b1 (1/d)**
- **h = b0 + b1 ln(d) + b2 (1/d)**

Equação de volume

Para realizar estimativas de volume das árvores das parcelas, foram ajustados os seguintes modelos:

- $v = b0 + b1 \cdot d^2.h$
- $v = b0 + b1 \cdot d + b2 \cdot h$
- $v = b0 + b1 \cdot d + b2 \cdot h$

Seleção de equações

A seleção da melhor equação de relação hipsométria e de volume foi realizada por meio das estatísticas:

- Coeficiente de Determinação ajustado (R^2aj)
- Erro Padrão de Estimativas em percentagem (Syx%)

RESULTADOS

A área da floresta de *Pinus cariboea* inventariada possui 5,0 ha. Os resultados do inventário são apresentados a seguir para a amostragem piloto e para a amostragem definitiva. O número potencial de unidades amostrais da população de 5,0 ha, com parcelas de 506,71 m^2 é de aproximadamente 99 unidades amostrais.

Cubagem e equações de volume

Foram cubadas 24 árvores distribuidas aleatóriamente na população, sendo os resultados do ajuste de equações de volume apresentado na Tabela 1, onde se observa que a equação de número 1 foi a melhor, com R^2aj de 0,98 e Syx% de 9%.

TABELA 1 – Resultados do ajuste de equações volumétricas para *Pinus cariboea* com 26 anos de idade em Bateri-AL

Nº	Modelo	Parâmetros			Estatísticas			
		b0	b1	b2	R^2	R^2aj	Syx	Syx%
1	$v = b0 + b1 \cdot d^2.h$	0,034046	4,02E-05	-	0,985367	0,984701	0,066214	9,00%
2	$v = b0 + b1 \cdot d + b2 \cdot h$	-0,97655	0,051537	0,018289	0,951633	0,947026	0,123212	16,76%
3	$v = b0 + b1 \cdot d + b2 \cdot h$	-0,97655	0,051537	0,018289	0,951633	0,947026	0,123212	16,76%

Onde: v = volume individual em m^3; d = diâmetro a 1,3 m de altura; b0, b1 e b2 = parâmetros ajustados; R^2 = coeficiente de determinação; R^2aj = coeficiente de determinação ajustado; Syx = erro padrão de estimativas em m^3; Syx% = erro padrão de estimativas em percentagem.

Relação hipsométrica

Nas 6 parcelas amostrais do inventário piloto foram medidos todos os diâmetros de todas as árvores e as alturas das 10 primeiras árvores de cada parcela para ajuste de equações hipsométricas. Os resultados do ajuste das equações é apresentado na Tabela 2, onde se percebe que a equação número 3 foi a melhor, com maior R^2aj (0,97) e menor Syx% (4,4%).

TABELA 2 – Resultados do ajuste de equações hipsométricas para *Pinus cariboea* com 26 anos de idade em Bateri-AL

Nº	Modelo	Parâmetros			Estatísticas			
		b0	b1	b2	R^2	R^2aj	Syx	Syx%
1	h = b0 + b1 ln(d)	-13,1285	9,402095	-	0,9581	0,9574	0,732	5,20
2	h = b0 + b1 (1/d)	23,1548	-153,229	-	0,9131	0,9116	1,054	7,49
3	h = b0 + b1 ln(d) + b2 (1/d)	-44,1805	17,38186	134,3939	0,9704	0,9693	0,621	4,41

Onde: h = altura em m; d = diâmetro a 1,3 m de altura; b0, b1 e b2 = parâmetros ajustados; R^2 = coeficiente de determinação; R^2aj = coeficiente de determinação ajustado; Syx = erro padrão de estimativas em m³; Syx% = erro padrão de estimativas em percentagem.

Inventário Piloto

No inventário piloto foram medidas 6 unidades amostrais com raio de 12,7 metros, com área de 506,71 m², correspondendo a uma área amostral total de 0,304 hectares, o que corresponde a 6,1% de intensidade amostral, induzindo a utilização das equações para população finita, pois a amostragem é superior a 2% da população.

Como resultado do inventário piloto, encontrou-se uma média de volume de 235,0 m³/ha com uma variância de 861,4 (m³/ha)². O erro máximo admitido é de 10%, ou 23,5 m³/ha.

Com estes dados estimou-se um número de 9 parcelas amostrais para obter-se uma média com erro máximo de +/-10%.

Inventário Definitivo

No inventário definitivo foram locadas no campo 10 unidades amostrais, entretanto a unidade de número 9 foi perdida, tendo-se substituído pela unidade de número 10. Foram medidas 9 parcelas das 99 unidades amostrais potenciais da população, representando uma intensidade amostral de 9,1%.

Os resultados por classe diamétrica são apresentados na Tabela 3 e os resultados por unidade amostral são apresentados na Tabela 4. A floresta apresenta um número médio de 728 árvores por hectare (N), diâmetro médio de 19,9 cm e altura média de 14,4 m. A área basal média é de 24,7 m²/ha e o volume médio é de 229,4 m³/ha, sendo que mais de 70% do volume está concentrado nas árvores com diâmetro acima do médio.

TABELA 3 - Resultados por classe de diâmetro do povoamento de *Pinus cariboea* em Bateri-AL

Classes de diâmetro (cm)				Freq	N	d	h	g	v	G	V	V
Ordem	>=LI	<LS	CC	(nº de árv.)	(árv/ha)	(cm)	(cm)	(cm)	(cm)	(m²/ha)	(m³/ha)	%
1	10,00	13,14	11,57	77	169	11,69	10,08	0,01083	0,0905	1,83	15,28	7%
2	13,14	16,29	14,71	30	66	14,90	11,82	0,01748	0,1402	1,15	9,23	4%
3	16,29	19,43	17,86	45	99	18,02	13,51	0,02556	0,2115	2,52	20,87	9%
4	19,43	22,57	21,00	52	114	20,96	15,16	0,03456	0,3029	3,94	34,54	15%
5	22,57	25,71	24,14	48	105	23,94	16,57	0,04505	0,4169	4,74	43,88	19%
6	25,71	28,86	27,29	62	136	26,97	18,08	0,05717	0,5643	7,77	76,72	33%
7	28,86	32,00	30,43	18	39	29,89	19,36	0,07023	0,7317	2,77	28,88	13%
Médias				332	728	19,90	14,43	0,03397	0,3151	24,7	229,4	100%

12

Coeficiente de Variação (%)	-	-	30,38	21,89	55,62	63,17	-	-	-

Onde: LI = limite inferior da classe de diâmetro em cm; LS = limite superior da classe de diâmetro; Freq = frequência de árvores na classe de diâmetro observada; N = número de árvores por hectare; d = diâmetro a 1,3 m de altura; h = altura em m; g = área basal média da classe em m²; v = volume médio da classe em m³; G = área basal por hectare em m²/ha; V = volume por hectare em m³/ha.

A Variância da população (s_x^2) foi de 633,4 (m³/ha)², resultando num Desvio Padrão (s_x^2) de 25,2 m³/ha e num Coeficiente de Variação de 11% (Tabela 4).

TABELA 4 - Resultados por parcela do povoamento de *Pinus caribaea* em Bateri-AL

Parcela nº	d (cm)	CV d (%)	h (m)	g (m²)	v (m³)	N (árv/ha)	G (m²/ha)	V (m³/ha)
1	18,1	32,7	13,5	0,02835	0,2573	730,2	20,70	187,9
2	19,9	30,2	14,4	0,03323	0,3020	789,4	26,23	238,4
3	20,7	27,0	14,9	0,03598	0,3312	690,7	24,86	228,8
4	18,8	34,2	13,8	0,03158	0,2985	749,9	23,68	223,9
5	20,2	39,9	14,6	0,03636	0,3514	769,7	27,99	270,5
6	20,4	29,6	14,7	0,03551	0,3300	789,4	28,03	260,5
7	19,6	28,2	14,3	0,03265	0,2993	710,5	23,20	212,7
8	22,1	24,0	15,6	0,03993	0,3709	572,3	22,85	212,3
10	19,8	25,6	14,4	0,03336	0,3063	749,9	25,02	229,7
Médias	20,0	-	14,5	0,03411	0,3163	728,0	24,7	229,4
CV %	5,7	-	4,2	9,7	10,6	9,2	9,8	11,0

Onde: d = diâmetro a 1,3 m de altura; CVd = coeficiente de variação dos diâmetros em %; h = altura em m; g = área basal média da classe em m²; v = volume médio da classe em m³; N = número de árvores por hectare; G = área basal por hectare em m²/ha; V = volume por hectare em m³/ha.

A Variância da Média ($S_{\bar{x}}^2$) ficou em 72,0 (m³/ha)² e o Erro Padrão da Média ($S_{\bar{x}}$) em 8,0 m³/ha.

Com uma Probabilidade de 0,05 e 8 graus de liberdade, o valor de t(p;gl) é de 2,306, que multiplicado pelo Erro Padrão da Média ($S_{\bar{x}}$) resultou num Erro Amostral Absoluto (Ea) de 18,4 m³/ha, ou um Erro Amostral Relativo (Er) de 8,0 %, estando abaixo dos 10% estabelecidos para a amostragem.

O intervalo de confiança (IC) para a média ficou em:

IC [211,0 m³/ha <= μ <= 247,8 m³/ha] = 95%

Portanto, a Estimativa Mínima de Confiança é de 211,0 m³/ha, ou de 1055 m³ para a população de 5 hectares.

13

CONCLUSÕES

Foi ajustada equação de relação hipsométrica com os dados do inventário piloto que apresentou alto Coeficiente de Determinação e baixo Erro Padrão de Estimativas, permitindo estimar as alturas das árvores das parcelas amostrais com boa precisão.

Foi ajustada equação de volume com os dados de 24 árvores cubadas com alto Coeficiente de Determinação e baixo Erro Padrão de Estimativas, permitindo estimar também os volumes das árvores das parcelas amostrais com boa precisão.

A distribuição diamétrica dos dados da população mostrou que os desbastes provocaram irregularidade da frequência nas classes de diâmetro, necessitando correções por meio de desbastes para se obter maior crescimento das árvores de maior dimensão.

O volume de estoque de madeira do povoamento de *Pinus caribaea* com 5 hectares foi determinado com erro amostral dentro do estabelecido para o inventário florestal.

RECOMENDAÇÕES

Recomenda-se realizar um desbaste por baixo na população com o objetivo de eliminar as menores árvores, principalmente as da menor classe diamétrica para permitir maior espaço para as árvores de maior porte que deverão ter seu crescimento intensificado com a liberação de espaço.

É recomendável realizar um inventário contínuo medindo-se as parcelas amostrais ao menos imediatamente antes dos desbastes e após os desbastes para determinar os critérios para os desbastes, determinar o quanto foi retirado e a situação após os desbastes bem como acompanhar crescimento e estabelecer a melhor maneira de conduzir o manejo do povoamento.

REFERÊNCIAS

FLORIANO, E. P. **Manejo florestal**: para sustentabilidade e excelência. Rio Largo: edição do autor, 2018.

COCHRAN, W.G. **Técnicas de Amostragem**. Rio de Janeiro: USAID / Fundo de Cultura, 1965.

Printed by Books on Demand GmbH, Norderstedt / Germany